The Atlantic Slave Trade History

Now in its second edition, *The Atlantic Slave Trade in World History* has been updated to include recent scholarship, and an analysis of how debates have changed in light of recent key events such as the Black Lives Matter movement.

Primarily focused on the Atlantic Slave Trade, this study places slavery within a broader world context and includes significant detailed coverage of Africa. With a chronological approach, it guides students through the origins of the Atlantic Slave Trade to its expansion and eventual abolition. Its final chapters explore the legacy of the Atlantic Slave Trade by comparing it to other systems of slavery outside of the Atlantic region, and analyze the persistence of modern-day slavery. As well as offering an analysis of historiography, the updated bibliography and conclusion, which considers the recent Black Lives Matter protests and their aftermath, provide a fresh account of how slavery has shaped our understanding of the modern world.

Unmatched in its breadth of information, chronological sweep, and geographical coverage, *The Atlantic Slave Trade in World History* is the most useful introductory resource for all students who study the Atlantic Slave Trade in a world context.

Jeremy Black is Emeritus Professor of History at the University of Exeter. He is the author of over 170 books, including *War in the Modern World, 1990–2014* and *Introduction to Global Military History*.

Themes in World History
Series editors: Peter N. Stearns and Jane Hooper

The *Themes in World History* series offers focused treatment of a range of human experiences and institutions in the world history context. The purpose is to provide serious, if brief, discussions of important topics as additions to textbook coverage and document collections. The treatments will allow students to probe particular facets of the human story in greater depth than textbook coverage allows, and to gain a fuller sense of historians' analytical methods and debates in the process. Each topic is handled over time – allowing discussions of changes and continuities. Each topic is assessed in terms of a range of different societies and religions – allowing comparisons of relevant similarities and differences. Each book in the series helps readers deal with world history in action, evaluating global contexts as they work through some of the key components of human society and human life.

The Turkic Peoples in World History
Joo-Yup Lee

Punishment in World History
Peter N. Stearns

The Environment in World History
(Second Edition)
Stephen Mosley

Globalization in World History
(Fourth Edition)
Peter N. Stearns

The Atlantic Slave Trade in World History
(Second Edition)
Jeremy Black

The Atlantic Slave Trade in World History

Second Edition

Jeremy Black

 Routledge
Taylor & Francis Group

NEW YORK AND LONDON

Designed cover image: Diagram of a slave ship from the Atlantic slave trade. From an Abstract of Evidence delivered before a select committee of the House of Commons in 1790 and 1791. Alamy.

Second edition published 2024
by Routledge
605 Third Avenue, New York, NY 10158

and by Routledge
4 Park Square, Milton Park, Abingdon, Oxon, OX14 4RN

Routledge is an imprint of the Taylor & Francis Group, an informa business

© 2024 Jeremy Black

First edition published by Routledge 2015

ISBN: 978-1-032-60173-1 (hbk)
ISBN: 978-1-032-59984-7 (pbk)
ISBN: 978-1-003-45792-3 (ebk)

DOI: 10.4324/9781003457923

Typeset in Times New Roman
by SPi Technologies India Pvt Ltd (Straive)

For Lester Crook

Contents

Preface viii
List of Abbreviations xiii

1 Introduction 1

2 The Beginnings of the Atlantic Slave Trade 21

3 The Slave Trade Expands Greatly 44

4 The Slave Trade at its Height 75

5 Abolitionism 119

6 After Slavery? 162

7 Conclusions 174

Selected Further Reading 189
Index 191

Preface

One can think that slavery belongs to the history books, but far from that, it's worse today than ever in history.
—Monique Villa, head of the Thomson Reuters Foundation, speaking at the Trust Women conference, London, December 3, 2013

This book addresses a key theme in world history, a theme that helped to make the modern world: the slave trade. This book seeks to establish the historical narrative of the slave trade, focusing on that across the Atlantic, from Africa to the Americas, from the fifteenth to the nineteenth centuries. In doing so, the book introduces important discussions about the causation of the slave trade, notably over the role and emergence of racism, over Africa's role in the trade, and on the causes of growing opposition to the trade and to slavery. There is also a discussion of what happened to the slave trade and slavery after their formal abolition in the nineteenth century.

The significance of the slave trade is many-sided. The human geography of the world, the distribution of peoples, and the nature of nations were transformed as a result of this trade, notably, but not only, in the Atlantic world. Moreover, the impact of the slave trade was seen in the character of the resulting societies. Indeed, this theme was brought to the fore in March 2014 when Caricom, the fifteen-strong Caribbean Community of states, demanded reparations from European slaving nations, including Britain and France, for the enduring legacy of the slave trade. The Caricom plan referred to "psychological trauma" as a consequence of slavery. As an instance of the varied legacy of the slave trade, the plan also called for the establishment of "repatriation programmes" to help resettle descendants of former slaves who wish to return to Africa, an idea with roots in the movement to abolish the slave trade and end slavery that began in Europe and North America in the late eighteenth century.

The slave trade has a long background, being well established in the Ancient World, and certainly from the third millennium BCE, but has been most significant for the development of the modern world since the fifteenth century, and that period is the focus of this book. In this period, the slave trade links to key themes: notably to globalization, the integration of different parts of the world; and to transnationalism, the extent to which new identities and interests across nations were created, particularly by migration and trade. This significance also leads to two crucial points. First, the slave trade was a disgusting aspect of human history and a stain on a large number of individuals, groups, societies, and, indeed, civilizations. An emotional response is necessarily part of our reaction as we are humans. Yet, it is less and more than what we should provide. We also owe those who suffered this terrible treatment an intellectual response that seeks to explain the hows and whys. Throughout, however, remember the human experience of the trade, the cruelty and suffering involved, and the extent to which these were central to deliberate and planned processes and procedures. Moreover, most of those who were enslaved remain silenced,[1] even though scholarship has helped illuminate the situation and notably so by highlighting the consequences for women who sometimes in past scholarship received insufficient attention.[2]

Second, the slave trade is history, but it is also part of the present world, and that, indeed, is a central aspect of the importance of the subject. Comparisons of severity and misery across history are problematic at best. This is not only because conditions are difficult to gauge and measure in any one period, let alone to contextualize and compare across time. There is always the issue of comparing what is not strictly comparable. How does work in a plantation in 1750 compare with indentured labor in 1900, or with being a woman held against her will and abused for prostitution in 2024? How, indeed, does work in a plantation in 1750 compare with work in a plantation in 1860? How does work in a plantation in 1860 compare with the situation for former slaves on the same land in the 1890s, with many of the former slaves controlled by vagrancy laws and other means? And how does this question vary if we are looking at Brazil, Jamaica, or Cuba, rather than America?

There are also the serious issues involved in making judgments across time, and notably so in the context of shifting values and the differing application of these values. Attitudes to the slave trade and slavery are changing precisely because it is now appreciated that they are not simply historical conditions but, rather, all too present today. Indeed, allowing for a degree of hyperbole in some claims about the scale of slavery in the contemporary world, it is, nevertheless, the case that the issue has become far more prominent on the world scale over the last decade and indeed over the last five years—and sufficiently so for the British government to announce in December 2013 that it would introduce legislation on the issue. Administrative and legislative measures in Britain reflect the growing concern about slavery. The deaths of exploited foreign cockle fishers in Morecambe Bay when the tide came in led, in 2005, to the establishment of the Gangmasters Licensing Authority, which, in 2013, secured a conviction of Audrias Morkunas who used physical threats and control over accommodation to force fellow Lithuanians to work for a pittance in Britain. The same year, a parliamentary bill to end modern slavery was revealed in draft. It included the appointment of an Anti-Slavery Commissioner to focus and sustain government action, and the imposition of a legal obligation to report all encounters with people who may have been trafficked or who are trapped in forced labor. Thus, there was an obligation on citizens to act against the modern slave trade and modern slavery.

In April 2014, at the international level, Pope Francis held a conference in the Vatican about slavery in the modern world. To mark it, Theresa May, then British Home Secretary, later Prime Minister, and Vincent Nichols, the Cardinal-Archbishop of Westminster, the most senior Catholic cleric in England, published a joint article under the title "We must all play a part in abolishing slavery." They wrote of the victims:

> Their complex needs, varied backgrounds and circumstances mean there is no "one size fits all" solution. Men, women and children: British and foreign nationals. Trafficked for cheap labour, into prostitution or domestic servitude, or forced into a life of crime … what they all tend to have in common is that they are socially and economically vulnerable, on the margins of society. Organised criminal gangs deliberately prey on and exploit people they perceive to have no voice.

Turning back to the speech cited at the outset by Monique Villa, a French journalist and women's rights advocate, she claimed that it cost as little as £60 to buy a human being, that there were about 30 million slaves in the world in 2013, and that:

> Most of us here today have probably met a modern-day slave without knowing it; it can be on a bus, in a nail salon, in a posh [up-market] hotel. Anywhere; in London, in New York, in Dubai. They walk among us.

Thirty million is more than the population of many countries—for example, Australia, which had a population of 26.5 million in 2023. The same point is true for many other countries, even if lower recent estimates of 20 or 21 million slaves are employed. Villa defined slavery as including human trafficking, forced and bonded labor, sexual exploitation, domestic servitude, and forced marriage, adding: "The common denominator of all these crimes is the evil intention to strip a human being of their freedom and then to use and abuse them, control and exploit them." The modern campaign focuses on a significant situation, one in which the position in traditional areas of slavery, such as West Africa, Sudan, and Pakistan, are joined by the brutal consequences of labor movements in the contemporary world. Labor movements is a bland term for the often highly dangerous means by which people travel seeking work or are trafficked to that end.

There is also the slavery arising from harsh political control. In December 2013, Amnesty International placed newspaper advertisements declaring:

At this instant in North Korea, as many as 100,000 helpless people are being subjected to cruelties so extreme that it is hard to believe it is really happening. These people are held in slave labour camps hidden in remote mountain wildernesses. Most will never be released. They will die of hunger, cold, exhaustion, disease, brutality, torture, or execution.[3]

It is unlikely, however, that, for a range of political and ideological factors, the modern campaign against slavery will have the same traction as its nineteenth-century forbear. Other campaigns—for example, on environmental issues—are more popular. At the same time, the greater prominence of slavery today than during the twentieth century provides a context within which to reexamine the situation in the past. So too does the realization that slavery is not simply a vestigial and marginal relic of discredited practices, but, instead, an aspect of continuing social activities and of dynamic global and regional economic links and social practices such as differential abortion rates in societies that put a premium on male children.

Such a reexamination of historical slavery and of the historical slave trade is encouraged by the contentiousness today of the situation in the past, and, more specifically, the question of how the past should be presented. The extent to which the slave trade can be used to serve as a more general litmus test of the acceptability of past societies is readily apparent. On January 5, 2014, in an interview with the *Sunday Times* (of London), Chiwetel Ejiofor, the lead actor in *Twelve Years a Slave*, a much-watched film about slavery in 1840s' America (by when Britain had abolished both the slave trade and slavery), demanded that British schools face up to the country's slave-trading past. The film's British director, Steve McQueen, who stated in 2014 that there were 21 million slaves worldwide, argued that slavery should be remembered in Britain in the same way as the Holocaust, the Nazi attempt to kill all Jews.

This comparison is highly controversial, not least because the purpose of slavery was to control and exploit, but not to slaughter, still less to destroy totally, an entire ethnic group. The comparison with the Holocaust is singularly inappropriate because the idea of 'inherited sin' or guilt is the essence of antisemitism, while those today who demand reparations or expressions of guilt and culpability today are suggesting that former slave-owning nations have inherited guilt.

At the same time, the very argument reflects the major importance of the slave trade and slavery in public discussion about the past and as key measures of cruelty and depravity. Both are frequently cited when seeking to establish moral standards for the present age and across time, and when discussing the concept of progress.

For this edition, I have profited much from the comments of Bill Gibson and four anonymous readers, from an invitation to take part in a debate at the University of Pennsylvania in 2023, and from recent opportunities from 2023 to visit and consider the history of slavery in Antigua, Barbados, Barbuda, Grenada, Guadeloupe, Martinique, Nevis, St Kitts, St Martin, and the Virgin Islands.

It is a great pleasure to dedicate this book to Lester Crook, an excellent editor who has greatly helped me in bringing my work to publication and provided a necessary calming role for many books.

Notes

1 S. White and T. Burnard (eds), *Hearing Enslaved Voices: African and Indian Slave Testimony in British and French America, 1700–1848* (Abingdon, 2020).
2 M.J. Fuentes, *Dispossessed Lives: Enslaved Women, Violence, and the Archive* (Philadelphia, PA, 2016); T. Nunley, *At the Threshold of Liberty: Women, Slavery, and Shifting Identities in Washington DC* (Chapel Hill, NC, 2021); C. Walker, *Jamaica Ladies: Female Slaveholders and the Creation of Britain's Atlantic Empire* (Chapel Hill, NC, 2020); M.P. Candido and A. Jones (eds), *African Women in the Atlantic World: Property, Vulnerability and Mobility, 1660–1880* (Woodbridge, 2019).
3 *Observer*, December 15, 2013.

Abbreviations

Add	Additional manuscripts
AM	Archives de la Marine
AN	Paris, Archives Nationales
BL	London, British Library, Department of Manuscripts
EcHR	Economic History Review
FO	Foreign Office
NA	London, National Archives

1 Introduction

Slavery is correctly one of the most emotive issues in history. Slavery was, and is, a vile institution, and a degrading and onerous situation. Enslavement and the trading and transport of slaves were brutal and demeaning. However, slavery has also been a constant for much, and possibly all, of human history, and has been practiced by many civilizations and societies. As a result, the effects of slavery reach to the present day, as discussed in the Preface. Nowhere more do the effects reach than in the distribution of peoples and in the continued relevance of slavery for modern debates over racism and its impact. This is notably the case in multiracial societies, the number and range of which greatly increased as a result of the slave trade.[1] The role of slavery in debates underlines the extent to which history indeed is the account we offer today of the past, and therefore relates, deliberately or not, directly or not, to present concerns.

These concerns help explain the concentration in this book on the Atlantic slave trade. This trade was very significant for the composition and culture of the modern population of America, the West Indies, and parts of Latin America, notably Brazil, as well as greatly affecting Africa. It is a trade that was important to the economics of the Atlantic world from the fifteenth to the nineteenth centuries, and a trade that was crucial to power and social dynamics in this world. Power is about people as much as territory; in many cases, more so, because land is given meaning and value by the people who settle and work it. The slave trade was also a traumatic activity that threw much light on the goals, methods, and ethos of the major European powers that dominated the trade, not least in eighteenth-century Britain. Thus, the slave trade offers a way to consider Western imperialism, power, and culture— indeed, Western civilization. It is particularly significant for Britain and America.

The slave trade from Africa transformed the demographics, economics, society, and politics of the eastern seaboard of the New World. About 12.5 million enslaved people were embarked from Africa to the Americas, although deaths in the Middle Passage across the Atlantic reduced the

DOI: 10.4324/9781003457923-1

number of the enslaved who arrived to about 10.7 million.[2] As an instance of the importance of the language chosen, I originally wrote "embarked," not "were embarked," but the former is totally inappropriate as it suggests a degree of choice as opposed to the reality of being treated as goods. "Deaths in passage," moreover, scarcely captures the horror of what happened, nor the degree to which the slave trade was responsible for the deaths.

The use of force to despoil or overawe native peoples had been a factor in European conduct in the Americas from the outset, but the African slave trade and slavery added a crucial new dimension of coercion, one on a great scale, and also added a new economic and racial geography. Thereafter, slavery could not be divorced from the geopolitics of the New World and the Atlantic. Indeed, the prominent role that slavery played in divisions within America from the 1820s, helping both to cement a sense of separate and particular Southern identity, and to give it an expansionist dynamic, was inherent to the geopolitics and development of North America. A key feature in the fate of the American state until the 1860s, slavery, principally through its strong regional and ethnic after-echoes, then played a major part in American politics into the twenty-first century.

The concentration on the Atlantic slave trade, however, can lead to a neglect of other such slave trades. We shall focus on these trades, before moving on to consider the genesis of the Atlantic slave trade. Because of the significance of these other trades, it was (and is) the case of multiple slave trades, rather than the Slave Trade. Indeed, the latter term is a misnomer if it implies that there was one essential character for the trade in slaves: that was never the case. What the slave trade means for the history of East Africa or the Mediterranean lands is different from what it means for the Atlantic world.

Definitions

As another indication of variety, we need foremost to consider definitions, for varied definitions of slavery are significant for the account of the slave trade. Slavery is similar to war: in one light, enforced servitude, like large-scale, violent conflict, is easy to define. You know what you see. However, just as discussion of war frequently overlaps with other aspects of conflict and violence—for example, rebellion and terrorism, let alone wars against poverty, crime, ignorance, or drugs—so the same is true with slavery, with force and servitude being open to varying definitions. The International Convention with the Object of Securing the Abolition of Slavery and the Slave Trade (an agreement that was ratified in 1926 by the members of the League of Nations, an international organization, the predecessor of the United Nations, that, due to the decision of Congress, did not include America), defined slavery as "the status or condition of a person over whom any or all of the powers attaching to the right of ownership" are exercised.

That might seem a clear definition, but, in practice, it only emerged after debate and political bargaining designed to protect vested interests and cultural practices. This bargaining led to the exclusion from the definition of slavery of forced labor and concubinage, both of which involved many people in conditions of slavery. The debate can be followed through the sources; and an understanding of this process of negotiation calls into question any attempt to present a definition of slavery as of universal use. For example, in 2000, the International Association Against Slavery decided to include debt bondage, forced work, forced prostitution, and forced marriage in the scope of slavery. This definition was a major extension to that of 1926, notably in the case of forced marriage.

There is also the question of work. It is not necessary to draw attention to the modern, often highly misleading term "wage slave," in order to note that many who are not formally seen as slaves have had little or no choice about work and its character and context, not least in terms of subservience and remuneration. Such a point, which affects discussion of "slave conditions" today, has for long been made. In the nineteenth century, for example, during the heated and lengthy debate in America over slavery, comparisons were drawn, notably, but not only, by those seeking to extenuate black slavery, between the black slaves in the American South and the white workers in many Northern company towns. The latter, in the eyes of critics, were made subservient to their employers by various means, including being paid in tokens redeemable only in company shops. Legal freedom thus appeared less important than economic freedom. This was an issue that both Marxists criticizing industrial capitalism and Southern apologists defending slavery could see as pertinent.[3] Subsequently, employment practices involving a marked degree of control could be referred to as slavery—for example, the contracts of the mid-twentieth century under which Hollywood stars were bound to work at the behest of the studios.

However, in practice, there was a major difference between legal and economic freedom, in that free laborers generally did not have to face the threats of physical abuse and of separation from family that slaves frequently confronted.[4] As a result of this element, slavery can, more readily, be compared not to work in nineteenth-century industrial cities, but to the serfdom seen in medieval Europe and also with many East European peasants in the sixteenth to nineteenth centuries.[5] The character of Russian serfdom was bitterly criticized in Alexander Radishchev's book *Journey from St Petersburg to Moscow* (1790), which denounced arduous work, poor living conditions, and the right of lords to sell and to flog serfs. These were all themes also taken up in Abolitionist literature in Britain and America in the late eighteenth and nineteenth centuries directed against slavery.

The slave trade might appear to open up an important distinction between slavery and serfdom—namely, the compulsory movement for work seen with slaves. However, those who were not slaves included many people subject to such movement—for example, transported convicts, others sent to colonies or into internal exile against their will, and even, in one light, the indentured servants and others traveling for economic opportunity within a system in which their choices were limited or nonexistent. To take an example from the history of England at a time when it was developing slave colonies in the West Indies, the suppression of large-scale opposition in Ireland in 1649–52 by the English army under Oliver Cromwell was accompanied by the dispatch of some prisoners—for example, from the town of Drogheda in 1649, to work the sugar plantations of Barbados in the West Indies. In addition, Royalist conspirators in England in the 1650s were sent to work in Barbados. Moreover, Britain went on transporting convicts to Australia until 1868, long after it had ended the British slave trade in 1807. These convicts were put to work—for example, building roads, bridges, and public buildings. Convicts who were not transported abroad could also be used for labor gangs.

Other Western states followed similar policies. Spain for example deporting prisoners from Mexico to the Philippines. In Spain's major Caribbean colony, Cuba, in the 1830s, those working on the railways included convicts transported from Spain, as well as captured Carlist rebels. Portugal sent penal exiles to its African colonies, notably to Luanda in Angola, São Tomé and the Cape Verde Islands.[6]

Problems in discussing the movement of people extend to the present. For example, people smuggling is not the same as trafficking because the latter creates a dependency on the part of those trafficked, and, in order to repay this dependency, they have to work, and usually in onerous conditions. There are also the issues of definition and comparison posed by serfdom. In 1776, the influential Scottish economist Adam Smith saw serfdom as a "milder kind" of slavery,[7] but many serfs would not have appreciated his distinction.

Complex issues of definition and debates over explanation, with the problems of judgment that thereby follow, are not only pertinent for the Western world, but also arise for non-Western societies.[8] For example, the China of the Shang dynasty (1766–1122 BCE) has been called a slave society. However, the contrary has also been argued on the grounds that most of the population were not bought or sold, nor were they deprived of their personal freedom, although they were subject to coercive work.[9] Similar points could be made about collective farms under Joseph Stalin, the Communist dictator in the Soviet Union from 1924 to 1953. Nevertheless, in the Soviet *gulags* or labor camps, the situation was much more similar to slavery.

Discussion of how far serfdom, or other forms of labor or life, constitute slavery therefore has to take note of the extent to which slavery itself was, and is, not a fixed category. Instead, slavery is open to a variety of definitions.[10] Most significantly, in advancing a typology of slavery, it is possible to differentiate between societies with slaves, in which slavery, while sanctioned by society, was largely a domestic institution providing labor in the household, and slave societies, in which slavery was the mode of production on which the dominant group depended for its position. It is also possible to focus on two types of the latter: slavery at the disposal of the state, and slavery within a private enterprise system. There was an overlap, not least because, even if readily contrasted categories, individual slaves could move from one to the other. In the Roman world, the state created slaves from defeated peoples, both noncombatants and serving troops, who could then be moved into the domestic sphere of usage within the empire. Yet, there could also be significant differences in the condition and treatment of slaves. Moreover, slavery at the disposal of the state tends to receive less attention than slavery within a private enterprise system, largely because the latter was the dominant type in the Atlantic world.

However, state slaves of various types were important in many pre-modern countries. There are numerous examples, from Antiquity to the end of the Chinese Civil War in 1949, where a defeated nation or side could be obliged to provide troops for the victor as part of the peace settlement, in effect producing slave army units. In some cases, state slaves were key elements in the governmental system, most obviously with the *janissary* units in the Ottoman (Turkish) army who played a crucial role in the army and the politics of the state until 1826 when, as an aspect of modernization, they were suppressed by Sultan Mahmud II. The Ottomans' "slaves of the Sublime Porte," the term used for the government, were maintained by a system of *devsirme* (collection) from non-Christian households, a slave system of considerable sophistication, with boys conscripted in the empire. Their training was described in *The Laws of the Janissaries*, which was written in the early seventeenth century. Boys were regarded as easier to train and subdue than men. The best-looking were allocated to the palace to receive an education in the palace schools and serve the person of the sultan, while the physically strong were chosen to work in the palace gardens, but most were destined for the military. First, they were assigned to Turkish farmers in Anatolia in order to accustom the boys to hardship and physical labor, and to teach them some basics of Islam and, more significantly, Turkish, which was the language of the *janissary* corps and the imperial elite.

Recalled after about seven or eight years, and based in the barracks of the novices near the palace entrance, the novices were used for palace and imperial tasks, such as transporting firewood to the palace or manning troop ferries, or working as palace laundrymen and apprentices in the naval dockyards, and on construction projects. The circumcision of the boys when they arrived for the first time at Constantinople was designed to assert their Muslim future and was a clear instance of symbolic power over the slave soldiers-to-be.[11]

Slaves were also important in the Ottoman navy. The galleys of Mediterranean navies depended on slaves, and they were frequently shackled to the oars. As a result, when galleys sank, as many Ottoman ones did in the battle of Lepanto in 1571, the slaves drowned. As a sign of their position, Ottoman galley slaves wore an iron ring on one foot. As with slave life on Western plantations in the New World, conditions for the galley slaves ranged from the brutal to a hardworking life, not too different from that of the bulk of the population. Galley slaves were kept in *bagni* quarters where, as in slave plantations in the New World, they were both under discipline and able to pursue opportunities to earn money. Again, as in the New World, some slaves were promoted to positions of authority, while others faced sexual exploitation, in this case homosexual rape.

Slave soldiers were important in other Islamic societies—for example, Morocco, Persia (Iran), the sultanate of Delhi in northern India, Achin in Sumatra, and the Islamic states of sub-Saharan Africa and Spain. The Mamluks, who ran an Egyptian-based empire from 1250 to 1517 that extended to include modern Syria, Lebanon, Israel, and western Saudi Arabia, were largely Circassians from the Caucasus, captured in childhood, enslaved, and trained as slave soldiers. The original source of many of the slaves was the wide-ranging Mongol conquests of the early thirteenth century, as their campaigning ensured the availability of large numbers of slaves who were shipped, often by Genoese merchants from their Crimean base at Kaffa, to the affluent market of Egypt. There was subsequently a racial hierarchy among the slaves there, with an elite of slaves from the Caucasus, while the bulk of manual slave work was carried out by slaves from sub-Saharan Africa, although some fought in the army. There was a religious dimension to slavery in the Islamic world—for example, in the eagerness with which Sunnis, both from Central Asia and from the Middle East, enslaved Shi'ites.

The concept of state or public slavery can be expanded to consider entire populations. Certain modern governments, most obviously North Korea, claim so much authority, wield so much power, and deny so many freedoms to the people, including that of movement, that their entire population can be regarded as slaves. This point is of wider historical relevance. Looking back to the example of Classical Greek writers commenting on the Persians and other Asian peoples, European political rhetoric in the early modern period also employed the juxtaposition, as contrast, of liberty and slavery, typecasting the subjects of political systems judged unacceptable as slaves. Thus, the *London Chronicle* of May 7, 1757 referred to "Absolute governments, where the Crown and the Church unite to make men slaves," adding that in Poland, where the aristocracy were powerful "the Populace [was] a Race of Slaves." This argument was used in particular by the British against the French, notably in the wars of the eighteenth century. Similarly, the argument was applied against the Jacobites, the supporters of the exiled

Stuarts who looked to France and supported France. On March 19, 1748, in the newspaper the *Jacobite's Journal*, Henry Fielding wrote of the Jacobites and their "execrable designs in favour of popery and slavery."

When the British joined commercial expansion with the liberty promoted by their government in their understanding of the character of their empire, they were not thinking of the slaves they themselves transported and controlled. Instead, they treated liberty or its absence as affecting Western peoples. So also with the use of the contrast between liberty and slavery by the American Patriots who were opposed to British rule at the time of the American Revolution (1775–83). There was a powerful theme of religious rebirth in such arguments, notably with reference to Moses leading the Jews out of slavery in Ancient Egypt into the "promised land" of Israel. In the eighteenth century, both the British and the Americans saw this episode as directly relevant to themselves, as with Handel's operetta *Israel in Egypt*.

Radicals who criticized their own system of government, or even just its position and policies, also repeatedly made reference to slavery. This was particularly the case in Britain, and this tendency was to affect American critics of the rule of George III. In 1742, in Britain, a pamphlet supposedly written by a West Country clothier to one of the MPs on the secret parliamentary committee established to investigate the recently fallen Walpole ministry offered a stark choice, with slavery again proving a key term: "You, Sir, are one of our National Jury; this the last stake for British liberty; the only method that can revive and confirm our ancient constitution, or leave on us the indelible trace of slavery for ever." Slavery had so central a place in British public discourse that it had also been cited by government supporters as a reason not to establish this committee. Speaking in Parliament, Henry Fox, a Whig placeholder, pressed the need for national unity:

> when I reflect that Europe is now engaged in a war [the War of the Austrian Succession in which France had attacked Austria in 1741], upon the event of which its liberty depends, and by which it will probably be enslaved, unless we interpose with the utmost of our strength, I cannot but be of opinion … that union amongst ourselves was never more necessary than it is at this important crisis.

The description, and therefore criticism, of undesirable circumstances within Britain in terms of slavery continued into the nineteenth century. Speaking after the abolition of the slave trade by the British government, but while slavery was still legal in its colonies, Arthur Thistlewood, the head of the unsuccessful Cato Street conspiracy to murder the British Cabinet at dinner and to seize power, declared, in 1820, at the end of his trial: "Albion [England] is still in the chains of slavery. I quit it without regret." He was, indeed, hanged. Such rhetoric contributed to Abolitionism (the movement against the slave trade and then slavery) by making it seem a goal for radicals and also desirable for liberals. Subsequently, such rhetoric itself frequently drew on Abolitionism, arguing that freedom from slavery had to be extended to unwelcome political circumstances in the country in question.

However, a characteristic of Western slavery was, in fact, that it was not a description of the system of government and certainly not within Europe, but, instead, predominantly part of the commercial economy. In addition, such slavery was overwhelmingly practiced in the Western world in colonies outside Europe, and particularly by the early nineteenth century. Slavery in the Western world was a system of servitude driven essentially by what is termed, without any irony, free enterprise. This situation provided the crucial context for the Atlantic slave trade: it was a response to economic need, and a product of the competitive search for economic opportunity and profit. Moreover, need and opportunity were served by economic means—a commercially organized and conducted trade.

Slavery and Racism

If the Western world, Europeans, their colonies, and their former colonies, were, as Africa, Asia, and the Americas all show, far from alone in the slave trade, it is also necessary to qualify the commonplace identification of slavery with racism. Historically, as also for Islamic societies,[12] although they were frequently linked, there was no necessary or inevitable relationship between slavery and racism, or, at least, racism as currently defined. This was seen in the ancient world with the enslavement of prisoners of war and convicts and went on being the case. Indeed, alongside racist constructions and economic explanations for slavery, there were in practice a range of reasons and forms. These extended not only to the "other," whether racial, religious, national, or more than one of the above, but also to members of one's own community deemed deserving of slavery.[13] The freedom granted to those of another race was part of this equation, with race itself not always a category, and, if it was, proving difficult to define.

In practice, the origins and early history of both slavery and the slave trade are unclear, and the "pre-history" of slavery is best approached from the perspective of anthropology. Slavery probably proved a means to structure society and to treat outsiders, ethnic or moral. Origin myths suggest such a pattern. Both such structuring and such treatment involved control: control as goal and control as means. There was likely to have been in some cases an overlap in the treatment of outsiders with that of animals. As such, racism was a key element in slavery, which became a response to the religious as well as racial "other"—in short, to different people. This process can also be seen in conflict between primates, with the defeated either killed or taken into the community of the victors.[14] As far as humans were concerned, captives lacked the family ties and connections they had enjoyed, and this powerlessness was a key aspect of their slavery.[15]

Control over people served to forward a variety of purposes, including household service, sex, and other forms of work. The development of large-scale agricultural systems and of mining for minerals greatly increased labor needs, and slavery was a central element of the ancient world where it also provided household service. As a result, there was significant demand for slaves and that at a time of low population. Slaves were available from internal sources, especially as criminals, but were also acquired from abroad.

Egypt, for example, obtained slaves from both warfare and trade. This was notably so in obtaining slaves from the south in modern Sudan.[16] This indicates the continuities involved in slavery, for such processes continued at a large scale in this region into the nineteenth century. Moreover, the current relationship between Sudan and both its region of Darfur and the now independent state of South Sudan can be understood as a further continuation. Trading for slaves, which the Egyptians of the Fifth Dynasty (2494–2345 BCE) certainly did from Punt (probably eastern Sudan or Eritrea), probably represented obtaining the spoils of war at second-hand.

This, indeed, was to be a characteristic of much of the slave trade across time, which underlines the extent to which the history of the trade is in part that of war and of the cruelty surrounding war. In the branch of military history known as war and society, it is necessary to appreciate that the attempts to create at a global scale separate categories and rights for prisoners of war and for civilians are essentially modern. This is particularly so in the case of prisoners from very different cultures, although not only in those cases. Aristotle (384–322 BCE), one of the two most influential Greek philosophers, criticized the practice of Greeks enslaving conquered Greeks. However, Aristotle's support, as natural, for the enslavement of "barbarians" and those with a limited rational faculty was to be influential among Christian jurists and commentators, and up until the nineteenth century.

The scale of the slave trade was considerable, reflecting the large-scale demand for labor and the plentiful opportunities from war. Conquering the city of New Carthage in modern Spain in 209 BCE during the bitter Second Punic War, the great Roman general Scipio Africanus turned the working men of the city into slaves. In the 160s BCE, another Roman general, Lucius Aemilius Paulus, the conqueror of Macedon, reportedly sold 150,000 people from the region of Epirus into slavery, while Julius Caesar (100–44 BCE) claimed to have sold tens of thousands of conquered Transalpine Celts (inhabitants of modern France) into slavery in 58–51 BCE.

The numbers of slaves passing through the great slave marts was formidable. The small Greek island of Delos in the Aegean Sea was made a free port in 166 BCE and, according to the geographer Strabo (*c*.64 BCE–23 CE), 10,000 slaves could be sold in a single day. In an example of the slave trader in turn punished, the capture of Delos in 88 BCE by Menophaneses, a general of Mithradates VI, King of Pontus, an opponent of Rome, was followed by the killing of many and the enslavement of its surviving population. This process was completed by a pirate attack on Delos in 69 BCE.

Such pirate attacks reflected the variety of forces involved in enslavement. Alongside states there were nonstate actors. Thus, when Alexander the Great of Macedon captured the major Phoenician port of Tyre in 332 BCE, which had mounted a strong defense, he enslaved those in the city who were not killed. The Phoenicians, traders based in the coastal cities of modern Lebanon, had acquired slaves from across the Mediterranean, notably North Africa and Spain. In turn, the Cicilian pirates at the southern end of the Taurus mountains in modern Turkey were responsible in the first century BCE for slavery and piracy in the eastern Mediterranean. This range throws light on the comment by St. Augustine of Hippo in his book *The City of God* (412–27), the fundamental work of Christian political thought, that there was scant difference between Alexander the Great and a band of robbers.

The far-flung nature of the conquests of the empires of Antiquity, notably Rome, produced slaves. The enslavement of convicted criminals also produced many slaves. In the first century BCE about 20 percent of the 200,000 strong population of Rome were slaves. Trade for slaves was another major source. In the case of Rome, as later for the Western powers and both Africa and the Americas, there was a range of relationships stretching from linking in with existing slave-trade networks, to adapting these networks, to creating new ones. In the first case, the Romans drew on networks that moved slaves from West Africa across the Sahara to North Africa, and from the Ethiopian Highlands and Yemen down the Red Sea to Egypt, and from southern Russia via the Black Sea. The parallels with the later Western slave trade with sub-Saharan Africa were many. One was the need to cooperate with African states. Thus, the Roman trade along the Red Sea involved cooperation with the kingdom of Axum in what is now Eritrea. Similarly, the slave trade from the Upper Nile required the cooperation of the kingdom of Cush in what is now Sudan.

As another parallel, notably with America in the nineteenth century, the importance of the slave trade from outside the Roman Empire declined with time, as the slave trade was increasingly maintained by reproduction among the slave population. In the case of Rome, this decline was related to the lack of expansion of the empire during its later history, resulting in fewer conquered people who could be sold by the state and its army into slavery. In contrast, in the case of America, this decline was due to the abolition of the slave trade and the extent to which slavery was not imposed on those conquered as America expanded to the Pacific at the expense of Native Americans and of Mexico.

In post-Roman medieval Europe, slavery was frequent until the twelfth century. Many slaves were obtained by raiding peoples of a similar racial background, albeit of a different identity.[17] For example, in Anglo-Saxon England (fifth to eleventh centuries), slavery was linked with nonmembership of the tribe, rather than the racism of recent centuries. There was slave-trading from Ireland, and anticipating the city's prominent role in African slave trade, many of these slaves were brought ashore at Bristol. Slaves were moved across the Irish Sea from west to east, for, once based in Dublin from the late ninth century, the Vikings, who had already raided the English coast across the North Sea from 793, more readily raided the coasts of England, Wales, and Scotland that were on the Irish Sea. DNA studies for the Icelandic population reveal Irish as well as Norse roots.

Slavery and Serfdom

Slavery in England declined because of a reduction in the availability of enslavable people after the royal house of Wessex established control over modern England in the mid-tenth century. As a reminder that economic and ideological factors were also important, this decline of slavery within England was linked to changing patterns of land use, particularly an increase in rented land, as well as to the influence of Christianity. The change in labor control was important: it was more economic to give slaves smallholdings so that they, instead, became servile tenants. This was often linked with a transition from slaves as single people to servile families. On the holdings, it was easier to support families. This ensured that the labor force reproduced itself, which was more useful for the landlord than purchasing slaves as those raised to slave labor were familiar with what it involved. The number of slaves in England probably declined from the early tenth century. Although they still formed a substantial group in the *Domesday Survey* of the country in 1086, there were few by the mid-twelfth century. For long, Church institutions had had slaves but this became much less common with the Church reform movement of the eleventh century.

Serfdom, instead, was the key form of labor control in England by the twelfth century, which is a reminder of the extent to which slavery was an alternative among a number of forms of labor control. Serfdom was a system of forced labor based on hereditary bondage to the land. Its purpose was to provide a fixed labor force, and the legal essence of it was a form of personal service to a lord, in exchange for the right to cultivate the soil. Serfdom was used to provide the mass labor force necessary for agriculture. It entailed restrictions on personal freedom that, in their most severe form, were akin to slavery, and "many aspects of medieval serfdom were very like slavery."[18] Serfs were subject to a variety of obligations, principally labor services. They also owed dues on a variety of occasions, including marriage and death. Serfs could also be sold. The extent to which serfdom should be seen as equivalent to slavery is contentious, notably because it appears to lessen attention to the distinctiveness and horror of the latter. Moreover, enslavement, separately, remained as a legal category in Europe, not least as a frequent penalty for criminal behavior.

Slavery and the "Other"

There were white slaves in the early modern period (sixteenth–eighteenth centuries), most obviously those who manned the oars of the large numbers of galleys that contested the Turkish advance in the Mediterranean. Alongside or despite this, there was a deeper identity than before of racialism and slavery. Enslavement was frequently the response to the "other": to other peoples (irrespective of their skin color), and other creatures. Indeed, the terms "slave" and *sclavus* recall the origins of many slaves in the Balkans, a source of slaves for both Christian Europe and the Muslim world. Treating conquered peoples and their offspring as slaves seemed as logical to many as treating animals, such as horses, as slaves. Horses and other animals, beasts of burden, were also the creation of God, and therefore part of the divine plan, but the fact that they could be readily subordinated and trained for service to humans apparently demonstrated a natural and necessary fate. In adopting these attitudes, Christian ideology overlapped with those of Greco-Roman Antiquity and of the Islamic world.[19] Like booty or commodities, slaves were also sold by the victor to cover the cost of war.

In a fictional form, the treatment of the "other" was captured in William Shakespeare's play *The Tempest* (1611), which, in part, drew on accounts of English transoceanic exploration and colonization, notably of the Atlantic island of Bermuda, discovered in 1609. In the play, Caliban, the inhabitant of the island, who has an accursed parentage (his father is the Devil, his mother a witch), is enslaved, in turn, by Prospero, a wise but exiled Italian ruler, and then, in response to his exposure to alcohol, by two drunken Italians. Called "thou poisonous slave" and "abhorred slave," Caliban is a coerced worker ordered to fetch in wood.

Conflict between cultures increased the possibility of seizing humans as booty, and also eased ideological and normative restrictions on enslavement. This was certainly seen in Christian conflict with Islam, with both sides enslaving captives—for example, during warfare between Christians and Moors in Spain from the eighth to the fifteenth centuries, and during the Crusades in the Middle East from the eleventh to thirteenth centuries. Similarly, the contemporaneous eastward expansion of Christendom in Europe in the Northern Crusades led to the expansion of labor control over non-Germans, notably on the southern and eastern shores of the Baltic, particularly Pomeranians, Wends, Estonians, Prussians, and Livs. Although the

Sards, the natives of the Mediterranean island of Sardinia, were Christian, they were regarded by the Pisans and Genoese as primitive, and many were enslaved by them in the eleventh century.

The difference between peoples could be noted by the law in asserting the legality of slavery. In the case of *Butts v. Penny* in 1677, the status of a slave was recognized in English law: "the Court held, the negroes being usually bought and sold among merchants, as merchandise, and also being infidels, there might be a property in them sufficient to maintain trover," a common law action to recover the value of personal property wrongly taken. This decision was also followed in *Gully v. Cleve* (1694), but, in *Smith v. Browne* (1701), Sir John Holt, the Chief Justice of King's Bench, a Whig (liberal politicians) and a reformer, declared that "as soon as a negro comes to England he becomes free; one may be a villein [serf] in England but not a slave," a point that was to be developed in Somerset's case in 1772. In 1707, in *Smith v. Gould*, Holt decided that "by the Common Law no man can have a property in another."[20]

The Range of African Slave Trades

The search for slaves was not restricted to Europeans. In the Arab world, the slave trade from Africa, both across the Sahara Desert and by sea, and across the Indian Ocean and the Red Sea, was more longstanding than the European trade in the Atlantic world. Although scholars are well aware of the Arab, Ottoman, and Indian Ocean dimensions to slavery, this knowledge is far less the case with the public debate, whether in Africa, America, or Western Europe. This trade does not fit with the narrative of Western colonial exploitation, and is therefore widely neglected in public history, as well as in the current demand for apology and compensation. So also with the development of the slave production of sugar cane in Europe: by the Arab rulers of Sicily, in the ninth to eleventh centuries, using Christian slaves.

This widespread neglect of the non-Western slave trades also cuts across the grain of a world history that should be sensitive to the relative significance of developments in the past. In the sixteenth century, the Portuguese are the focus of attention for Western[21] activity in Africa and Asia. However, Portugal, the first of the Western powers to reach the Indian Ocean, in fact made much less of an impact in India than the Mughal conquerors of north India. So also in West Africa, where, from the fifteenth century, the Portuguese were the first of the Western Europeans trading slaves from sub-Saharan Africa. Nevertheless, in many respects they had less of an impact than Moroccan expansionism. A Moroccan expeditionary force crossed the Sahara and in 1591, at the battle of Tondibi, used its musketry to defeat the cavalry of Songhai, overthrowing the Songhai Empire, the leading state of West Africa. Instead, the Moroccans created a Moroccan

Pashalik of Timbuktu, which helped strengthen a major route for trans-Saharan trade, including that in slaves.[22] The Moroccans had already demonstrated the folly of writing the history of Africa from the perspective of Western pressure, when they smashed a Portuguese invading force at Alcazarquivir in 1578, killing the vainglorious king, Sebastian, and ending the long-standing Portuguese attempt to establish a powerful position in Morocco. Another instance of Islamic pressure on sub-Saharan Africa was provided, further east, by Idris Aloma, *mai* (ruler) of Bornu (1569–*c.*1600), an Islamic state based in the region of Lake Chad, who obtained his musketeers from the port of Tripoli (in modern Libya) on the Mediterranean, which had been captured from the Knights of St. John by the Ottomans (Turks) in 1551.

Christendom, in short, was in retreat, and other active powers were involved in the slave trade. This repeated a longstanding pattern of Muslim slave raiding at the expense of Christendom. Thus, in 1535, there were about 20,000 Christian slaves in the port city of Tunis when it was successfully attacked by a Spanish expedition. In turn, the local population was enslaved. Nine years later, the island of Ischia off Naples was devastated by Algerian raiders when it refused to provide tribute in the shape of slaves and money.

Similarly, in the sixteenth century, the musketeers and cannon provided by the Ottoman conquerors of Egypt helped Ahmad ibn Ibrihim al-Ghazi in the 1520s–40s to sustain his *jihad* in the Horn of Africa (Ethiopia and Somalia) against Christian Ethiopia. This holy war produced Christian slaves. In turn, as a reminder of the variety of Western-African relations, Ethiopia was supported in the early 1540s by Portuguese troops. Indeed, the history both of Africa and of its relations with Europe looks very different from the perspective of the Horn of Africa as opposed to the usual emphasis on West Africa. This point is not made to minimize the suffering and impact of the Atlantic slave trade, which, indeed, had fundamental effects on Africa, the New World and European imperialism. Nevertheless, the point serves to emphasize that the terrible suffering of the Atlantic slave trade was not unique and that a history of slavery centered on Africa, rather than the Atlantic, requires a broader account than one restricted to the Atlantic slave trade or largely written about it.

In fact, a focus on Ethiopia in the sixteenth century offers a potent reminder that Westerners did not come to Africa simply as oppressors, and, in addition, there were many non-Westerners among the latter. Further west, the state of Bornu captured slaves by raiding and, aside from its own use, transported some slaves north across the Sahara to the well-established slave markets of North Africa such as Algiers, Tunis, Tozeur, Jerba, and Tripoli. To the east of Bornu, three other states developed in the *sahel* (savanna) belt between the Sahara and the forests further south: Baquirmi, Darfur, and Wadai. These states used their military strength to acquire slaves who they sent to the markets in North Africa, a trade that acknowledged the gradients of wealth and influence. Darfur went on oppressing the non-Muslims to the south, in what is now South Sudan, into the twentieth century, and this role was taken on, both within Darfur and more widely by Sudan after it gained independence in 1956.

The slave trade across the Sahara was different from that across the Atlantic for a number of reasons, including the role of Islam in the former. Furthermore, the prime demand in the Americas was generally for male labor to work in the plantations, which, as a result, ensured a sexual imbalance against women in local societies affected by the trade.[23] However, in Jamaica in the late eighteenth century, buyers preferred young women and children as much, in part because they were easier to control, and in part because of higher taxes on those over 25.[24] In the case of the trade across the Sahara, the demand was largely for women, particularly for domestic servants and as sex slaves. Although there was plantation agriculture in some areas, there was not the equivalent in the Islamic world to the significant plantation economy that was short of labor seen in the New World, and notably in Brazil, the West Indies, and the southeast of what became America.

In addition to the important slave trades across the Atlantic and the Sahara, there were other significant areas of slaving in Africa. East Africa was a major source of slaves. They were traded, by Arabs, across the Red Sea and, further south, across the Indian Ocean, to markets in the Middle East, especially in the Arabian peninsula. In the ninth and tenth centuries, the Abbasid caliphate of Baghdad, the center of the Islamic world, drew in part on African slaves, as well as on Turkic slaves from Central Asia, Slavs from Eastern Europe, and Western European slaves via traders in Prague, Venice, and Marseilles. Thus, the Islamic and Christian worlds were closely linked as part of an active slave-trading system.[25] The slaves from Africa crossed the Red Sea to Jeddah, and then came overland to Baghdad, while others came via the (Persian) Gulf and the great slave market at Basra. Moorish Spain used captured Christian soldiers as slaves—for example, for agricultural work in Andalusia and Granada. Slaves were also imported into Moorish Spain from the Balkans.

The significance of slavery for Africa, but also the variety of circumstances involved,[26] is indicated by Madagascar, the largest island in the Indian Ocean. In its center, the kingdom of Merina, which expanded in the eighteenth century, was given cohesion by a sacred monarchy, force by firearms, and purpose by warfare for slaves. This was a slave trade that, like so much of that trade, was part of the wider world of slave movements, but also distinctive. Madagascar was the major source of slaves for the plantations on the French-controlled islands in the Indian Ocean, Réunion and Mauritius, which were, at once, outliers of the Atlantic world and the key to

France's significant and profitable presence in the Indian Ocean. As in West Africa, the European territorial presence in Madagascar was very limited, with a French coastal base at Fort Dauphin from 1746 to 1768; and, as in West Africa, the trade was dependent on African cooperation. For example, Andrianampoinimerina, ruler of Ambohimanga (r. *c*.1783–*c*.1810) in the center of the island, used slaving to acquire guns and gunpowder from the coast where Europeans traded; he seized slaves from other Malagasy territories and exchanged them for these weapons. Having conquered part of the interior, he left his successors to complete the task, which produced more slaves. Slavery was part of the labor control that was an aspect of expansionism in Madagascar. Once conquests were made by his successors, rebellions provoked by demands for forced labor had to be suppressed. Madagascar was not to be conquered by the French until 1894–5.

The number of Africans traded across the Sahara, the Red Sea, and the Indian Ocean is difficult to estimate, far more so, for source reasons, than the Atlantic trade. However, in combination, the number in these trades was probably as numerous as in the Atlantic trade, and there are suggestions that it was more so.[27]

The Western Quest for Labor Control

The Atlantic slave trade was an aspect of the quest for labor for a widening Western economy. It was a quest that was made more necessary by the extent to which labor that could be enslaved or controlled was not obtained in the Americas by Western Europeans in sufficient quantities by conquest, but was bought in Africa and brought from there. This situation was in contrast with warfare across much of the world, not only in Africa, but also with the expansion of Asian polities. Thus, in Asia, the Mongols in the thirteenth century, the Mughals in the sixteenth and seventeenth, and the Manchus in the seventeenth acquired slaves by warfare in the areas they conquered. From the perspective of labor availability, Western states and merchants suffered in this quest for labor by the extent to which the norms of Western war did not allow for the enslavement of captives in legitimate warfare between Christian states. Instead, they had to look elsewhere.

Labor was a key economic requirement in a world that by modern standards was underpopulated. Coterminous with the establishment of Western slavery in the New World and throwing light on it, rural society in Eastern Europe was transformed in the sixteenth and seventeenth centuries towards a "second serfdom," with heavy labor services provided by the peasantry. The causes of both can be discussed in terms of land–labor ratios and the search for controlled labor.[28] Paralleling the role of plantation exports from the New World to Europe, this "second serfdom" was a response to the commercial opportunities of early-modern grain exports to other parts of Europe, notably the Mediterranean. This trade was an aspect, at the European scale, of what is termed "globalization." The "second serfdom" appears also to have been, at least, prefigured by fifteenth-century changes, as lords, who had gained private possession of public jurisdictions, responded to the economic problems of the late medieval period, particularly fixed cash incomes. Thus, labor control was a means to raise income. It also reflected the extent to which there was no legal equality between subjects. The slave trade was an aspect of a similar situation, one in which mobility was provided not only by the export of goods but also by the enforced import of labor. Paralleling the role of landlords in the New World, the attitudes and powers of landlords in Eastern Europe, not least the character of their seigneurial jurisdiction over the peasants, were crucial to the spread of serfdom. The state stood aside or stepped back, and peasant rights were lessened—for example, in the Russian legal code of 1649. In another parallel with New World slavery, ethnic divisions—for example, between German landlords and Polish peasants, or Polish landlords and Ruthenian and Ukrainian peasants, exacerbated differences in some areas, and was important to the character and practice of this serfdom. These divisions were accentuated by religious contrasts—for example, between Protestant German landlords and Catholic Polish peasants, or Catholic Polish peasants and Orthodox Ukrainian peasants. As another instructive parallel, serfdom ended in nineteenth-century Europe, at the same time as slavery came to an end in the Americas.

Western slavery thus represented an aspect, an extreme aspect, of the commodification of human beings for reasons of labor that is central to economic activity. That commodification commonly involved a high level of control, indeed coercion, and was harsh for much of human history. Focusing modern concerns, Western slavery also reflected particular socio-cultural assumptions and practices. In these, nationhood, religion, and, even more, ethnicity, all played important, although varying, roles. These assumptions both interacted and became operative in particular contexts. If racism helped create slavery, slavery did the same for racism.

A key context for the development of slavery in the modern world was to be Western transoceanic expansion. This expansion, however, was not to be the automatic motor of the development of Western-controlled slavery, and, therefore, of a slave trade to sustain it. For example, Western expansion also involved the establishment of bases and colonies in a number of areas, from Newfoundland to Java, in which slavery did not become the pattern. However, enforced labor that can be seen as akin to slavery could still be important, as with the Dutch plantation economy on Java from the seventeenth century, an economy that produced spices. The Russian treatment of the peoples of Siberia conquered from the late sixteenth century was also harsh, not least with the seizure of local women, who were traded, and with the forced tribute in furs, *yasak*. This variety underlines the extent to which

European commercial expansionism in itself, like labor exploitation, was not coterminous with slavery. Nevertheless, for reasons of economic requirement, racial attitudes, and opportunities, slavery was to be a central aspect to much of this expansionism in the Atlantic world.

The Origins of Western Expansion into Africa

The historical tradition of slavery that was to be directly relevant to the initial development of the Atlantic slave trade was that in Portugal and Spain. Among Western Europeans, the Portuguese and Spaniards had the longest experience of conflict with Islam as a result of the *Reconquista*, the process in which the Moorish invaders were resisted from the eighth century, and then driven back from the eleventh; the kingdom of Granada, the last Moorish territory, finally fell in 1492. This conflict, which was etched deeply on the Iberian (Portuguese and Spanish) consciousness, ensured a supply of Moorish slaves. For example, when the islands of Minorca and Ibiza in the Mediterranean were captured by Alfonso III of Aragon in 1287, much of the Muslim population was sold into slavery, and, as a result, dispersed across the Mediterranean. In 1310, when the island of Jerba off modern Tunisia unsuccessfully rebelled against rule by Aragon, three-quarters of the island's population was enslaved. This was part of a longstanding pattern in particular sites. Thus, in 1135, Roger II of Sicily had invaded Jerba, killing or enslaving much of the population. The Catalan merchants who supported Alfonso III sold slaves in such major slave markets as Majorca, Palermo, and Valencia.

Slaves were also obtained from the conflict with the Moors in north-west Africa that overlapped and followed on from the *Reconquista*. The city of Ceuta in Morocco fell to the Portuguese in 1415. As a result of raiding and trading (the latter frequently trading in slaves obtained by raiding), slaves from a range of sources could be obtained in slave markets. Thus, in the 1390s, the slave markets of Majorca and Palermo sold Berbers from North Africa as well as Circassians from the Black Sea.

The expansionism against the Moors was to provide a context within which the opportunities were grasped by the Portuguese and Spaniards for enslavement from sub-Saharan Africa. Some of the slaves were Muslims, for Islam had spread south of the Sahara, notably into Senegambia in the twelfth and thirteenth centuries.[29] Moreover, the territories of the West African Islamic state of Mali reached to the Atlantic coast. However, unlike the Moors, these Muslim victims, let alone slaves from non-Muslim peoples further south, were peoples with whom there was no traditional antipathy. Portugal led the way in acquiring African slaves in West Africa in the 1440s, but Castile, the foremost Spanish kingdom, followed from 1453, until, in

1479, by the Treaty of Alcáçovas, Castile surrendered claims to trading rights in Guinea and the Gold Coast in West Africa to Portugal. A major source for slaves had been established. The expansion of Europe's Atlantic world to include the Americas was eventually to add unprecedented demand for slaves.

Notes

1 C.A. Palmer, "Defining and Studying the Modern African Diaspora," *Perspectives*, 36:6 (1988), 22–5; B.C. McMillan, ed., *Captive Passage: The Transatlantic Slave Trade and the Making of the Americas* (Washington, DC, 2002).
2 D. Eltis and D. Richardson, *Atlas of the Transatlantic Slave Trade* (New Haven, CT, 2010), 89, 203.
3 C. Vann Woodward, *American Counterpoint: Slavery and Racism in the North–South Dialogue* (Boston, MA, 1971).
4 J. Roberts, *Slavery and the Enlightenment in the British Atlantic, 1750–1807* (New York, 2013), 291.
5 For the economy of serfdom, A. Kahan, *The Plow, the Hammer and the Knout: An Economic History of Eighteenth-Century Russia* (Chicago, IL, 1985). For slavery, R. Hellie, *Slavery in Russia, 1450–1725* (Chicago, IL, 1982).
6 E.M. Mehl, *Forced Migration in the Spanish Pacific World: From Mexico to the Philippines, 1765–1811* (Cambridge, 2016); T.J. Coates, *Convict Labor in the Portuguese Empire, 1740–1932: Redefining the Empire with Forced Labor and New Imperialism* (Leiden, 2014).
7 A.R. Ekrich, *Bound for America: The Transportation of British Convicts to the Colonies, 1718–1775* (Oxford, 1987); A. Smith, *An Inquiry into the Nature and Wealth of Nations* (1776; Oxford ed., 1979), 386.
8 J.C. Miller, *The Problem of Slavery as History: A Global Approach* (New Haven, CT, 2012); J. Flynn-Paul and D. Pargas (eds), *Slaving Zones: Cultural Identities, Ideologies, and Institutions in the Evolution of Global Slavery* (Leiden, 2018).
9 J.A.G. Roberts, *A History of China* (2nd ed., Basingstoke, 2006), 5.
10 N. Lenski and C.M. Cameron (eds), *What is a Slave Society? The Practice of Slavery in Global Perspective* (Cambridge, 2018).
11 C. Imber, *The Ottoman Empire* (2nd ed., Basingstoke, 2009), 123–7.
12 C. El Hamel, *Black Morocco: A History of Slavery, Race and Islam* (Cambridge, 2013).
13 M. Gausco, *Slaves and Englishmen: Human Bondage in the Early Modern Atlantic World* (Philadelphia, PA, 2014).
14 R. Wrangham and D. Peterson, *Demonic Males: Apes and the Origins of Human Violence* (Boston, MA, 1996).
15 J.F. Brooks, *Captives and Cousins: Slavery, Kinship, and Community in the Southwest Borderlands* (Chapel Hill, NC, 2002).
16 I. Shaw, "Egypt and the Outside World," in Shaw (ed.), *The Oxford History of Ancient Egypt* (Oxford, 2000), 322–4.
17 A. Rio, *Slavery After Rome, 500–1100* (Oxford, 2017).
18 A.L. Poole, *From Domesday Book to Magna Carta* (Oxford, 1955), 40.
19 J.R. Willis (ed.), *Slaves and Slavery in Muslim Africa. I. Islam and the Ideology of Enslavement* (London, 1985).

20 W.S. Holdsworth, *History of English Law*, VI (London, 1924), 264–5. See also J. Brown, *Slavery and Islam* (London, 2019) and B.K. Freamon, *Possessed by the Right Hand: The Problem of Slavery in Islamic Law and Muslim Cultures* (Leiden, 2019).

21 The term "Western" distinguishes Christian Europe from the Muslim Europe that was part of the Ottoman (Turkish) empire.

22 S. Jeppie and S.B. Diagne (eds), *The Meanings of Timbuktu* (Cape Town, 2008).

23 J. Thornton, "The Slave Trade in Eighteenth Century Angola: Effects on Demographic Structures," *Canadian Journal of African Studies*, 14 (1980), 417–27.

24 A. Diptee, *From Africa to Jamaica: The Making of an Atlantic Slave Society, 1775–1807* (Gainesville, FL, 2010).

25 M. Jankowiak, "Dirhams for Slaves: An Early Medieval Slave Trade System," *The Oxford Historian*, 11 (2014), 26–30.

26 P.E. Lovejoy, *Slavery in the Global Diaspora of Africa* (Abingdon, 2019).

27 R.A. Austen, "The 19th Century Islamic Slave Trade from East Africa: A Tentative Census," in *The Economics of the Indian Ocean Slave Trade in the Nineteenth Century*, ed. W.G. Clarence-Smith (London, 1989), 21–44.

28 E.D. Domar, "The Causes of Slavery or Serfdom: A Hypothesis," S. Engerman, "Some Considerations Relating to Property Rights in Man," *Journal of Economic History*, 30 (1970), 18–32, 33 (1973), 43–65.

29 M.A. Gomez, *Black Crescent: The Experience and Legacy of African Muslims in the Americas* (New York, 2005).

2 The Beginnings of the Atlantic Slave Trade

Born in the Bissagos Islands off West Africa, Benkos Biohó was seized by a Portuguese slave trader and sold to a Spaniard at Cartagena, escaping in 1599 into the nearby marshes and came with other Maroons or escaped slaves to dominate the nearby Montes de María. The encouragement of other escapes posed a threat to Spain but, after a failed attempt to defeat the Maroons, peace was negotiated in 1612, only to be treacherously broken in 1619 when Biohó was seized. He was executed in 1621, but the Maroons in the Montes de María continued their defiance. There is a modern statue to Biohó in San Basilio de Palenque, a village founded by him as the first free African town in the Americas. It is the only one of these walled settlements founded by escaped slaves to survive in modern Colombia. The present inhabitants are the descendants of the slaves and preserve Palenquero, a Creole language, as well as African customs.

From start to end, Atlantic slavery and the Atlantic slave trade were not add-ons to Western development and civilization. Instead, they were linked with some of their major features and each should be considered in the light of the other. In some cases, the links were close, notably between the Age of Discovery and the development of the trade, and the role of the Age of Revolution in challenging it. At other times, the links were less close—for example, between the Scientific Revolution of the seventeenth century and the mercantilistic thought that encouraged the further growth of the trade. Throughout, it is necessary to put the slave trade and slavery in this wider context, while also using them to interrogate aspects of Western development.

In the late fifteenth and sixteenth centuries, the slave trade was an important part of a world of expanding empires and developing trade routes. This was not a world in which the Western powers were the only ones to develop empires. Indeed, prior to 1492, the most impressive on the global scale was the expanding Ottoman (Turkish) Empire. Moreover, even after Columbus laid claim for the rulers of Spain to new territories in the West Indies in 1492, the further expansion of the Ottoman empire had more of an impact in Eurasia—notably, its conquest within the world of Islam of the Mamluke

DOI: 10.4324/9781003457923-2

Empire of Egypt and Syria in 1516–17 and in 1534 of Iraq, and, at the expense of Christendom a rapid range of conquests including Belgrade in 1521, Hungary from 1526, and Cyprus in 1570–1.

Nevertheless, the creation of new routes of trade and power across the Atlantic transformed the geopolitics of Eurasia. The slave trade was linked to this geopolitics, but geopolitics was not the sole factor in the development of the trade. The interaction of economics and demographics, of demand, supply, and labor, were crucial. Western slavery was focused on the Atlantic world and not the Indian Ocean, the other major area of Western expansion from the cusp of the fifteenth and sixteenth centuries. In essence, this contrast reflected the key role of economic need, although opportunity was also an important factor. In the Indian Ocean world and its outliers, such as the (Persian) Gulf and the South China Sea, the Westerners did not require large quantities of slave labor and, anyway, despite some slave raiding and purchase, did not have the military means to provide such labor by force.

Western plantation economies did, it is true, develop in this far-flung Indian Ocean world, especially, under the Dutch, in parts of the East Indies (modern Indonesia), and some of these economies, such as the French island of Mauritius with its sugar economy, used slaves. However, most of the goods the Westerners brought back to Europe from the Indian Ocean around the Cape of Good Hope were not produced by slaves and, indeed, were obtained by purchase, notably from China. The Dutch, English, and French East India companies all played a key role in this trade. Furthermore, in the case, for example, of tea from China, there was no way in which this relationship could be altered, as, despite occasional fantasies of conquest, the Westerners were in no position to dictate to China. As a result, the tea trade was very different from that in coffee.

Coffee production, in contrast, was developed from the seventeenth century in the slave plantation colonies of the West Indies, notably the French-ruled islands of Guadeloupe and Martinique. Coffee was produced using slave labor. There was not enough cheap labor to do so by any other means.

The ability of the Westerners to establish plantation economies, and to move from trading bases to colonies, was limited around the Indian Ocean and, even more, in the Orient. This point was to be demonstrated further when fortified bases that had been established by Western powers could not be defended successfully from non-Western attack. This process was seen with the Dutch base of Fort Zeelandia on Taiwan in 1662, and the Portuguese bases on the Kanara coast of India in the 1650s, and of Fort Jesus at Mombasa in 1698. The English bases in India proved highly vulnerable to pressure from the Mughal Empire in the late seventeenth century. These and other examples demonstrate the extent to which the argument that there was a military revolution greatly increasing relative Western power needs serious qualification.

Spain and Portugal

The situation was different in the Atlantic world. There was an opportunity to establish plantation economies under Western control. In addition, to that end, labor was needed and labor was available, but, crucially, not at the same place. The need for labor sprang from the inherent demographic differ-ence between the Americas[1] and the far more populous South Asia, from the impact of Western expansion following Christopher Columbus's first voyage in 1492, and from the specific labor tasks that the colonists required.

Moreover, Spain and Portugal had developed methods of slave exploita-tion and trading in the Canaries and Madeira, islands in the Atlantic off the west coast of Africa. Portuguese settlement in Madeira began in 1424, and it became a leading producer of sugar using slave labor, with the slaves obtained from West Africa. This was the origin of the Atlantic slave trade.

The earlier focus of slave trading in Europe had been on the Mediterranean and the Black Sea, although, within this focus, there were frequent changes. In particular, the overthrow of the Byzantine (Eastern Roman) Empire by the Fourth Crusade in 1204 had been followed by the large-scale establish-ment of traders from the Italian peninsula, notably Genoese and Venetians, in the trade of the region. This included the existing slave-trading economy based on the Black Sea. The cities of Kaffa in Crimea and Trebizond on the coast of modern Turkey were the key markets for trade, and, through these, the Genoese and Venetians drew on long-established trade routes in Eurasia, and also provided slaves to Mediterranean markets, both Christian and Muslim. It was also along these routes, and through Kaffa, that the Black Death (bubonic plague) was to reach Europe in the 1340s. The khan-ate of the Golden Horde, one of the successor states to the Mongol Empire, was the key power in the region to the north-east of the Black Sea. It pressed on the Russian principalities, such as Kiev, Vladimir, and Galich, and its raiding produced large numbers of slaves. They were then moved to the Black Sea via the lower Volga and Crimea, both of which were part of the khanate.

However, as a demonstration of the central role of power in shaping the slave trade, the advance of the power of the Ottoman (Turkish) empire greatly changed the movement of slaves of all races in the fifteenth century. The Ottoman conquest of Constantinople (modern Istanbul) in 1453 was fol-lowed by the transformation of trading networks. The slave trade, notably that from the Black Sea, was focused on the new Ottoman capital and Ottoman demand replaced that from Venice and Genoa. Moreover, in the Mediterranean, the direction of expansion changed in the second quarter of the sixteenth century, with the Spaniards being pushed back in North Africa by the Ottomans. This greatly reduced the flow of slaves to Western Europe from North Africa, both Moorish slaves and those brought across the Sahara.

The need for new sources of supply for Christian Europe encouraged the focus on obtaining slaves from West Africa. Indeed, in the fifteenth century, Venetian and Genoese commercial, financial, and shipping interests that had been significant for the Black Sea slave trade transferred their expertise and capital to the new Portuguese-controlled African slave trade, as well, subsequently, as to Spanish expansion into the Americas. In this, and other ways, a range of developments and needs came together in the expansion of the slave trade.

The impact of disease rapidly and catastrophically brought by the Western invaders of the Americas was also highly significant in encouraging the expansion of the slave trade from Africa. This impact was particularly true of smallpox, which broke out in Mexico in 1520, the year after a Spanish invasion force under Hernán Cortés arrived. Smallpox appears to have killed at least half the Aztecs of modern Mexico, including their energetic leader, Cuitlahuac, and to have hit the morale of the survivors. Disease weakened potential resistance to Western control in the Americas. It acted like enslavement in disrupting social structures, and household and communal economics, leading to famine. From the Western perspective, disease also savagely hit the potential labor force. Infectious disease was a great killer, as it was later to be for slaves transported across the Atlantic.

The practice of moving natives to work, which exacerbated the situation, had already been practiced by Spain in the Canaries. Smallpox decimated the population of Hispaniola, the island—subsequently divided between Saint-Dominque, now Haiti, and Santo Domingo, now the Dominican Republic—which is the second biggest in the Caribbean after Cuba. More generally, Spanish and Portuguese colonial policies and practices, including the end of native religious rituals, demoralized the native population and limited the possibility of post-epidemic demographic (population) recovery.[2]

This situation created acute problems as the colonists wished to exploit their new possessions. To do so required labor, and lots of it, and notably as there were relatively few colonists.

The potential labor force available to the Westerners in the Americas was also limited by native resistance. This was to be important in British and French North America from the seventeenth century, but was already apparent in the sixteenth century in what became Latin America. From the outset, indeed, the Spaniards had encountered difficulties when expanding, although they were used from expansionist conflict against the Moors in Spain to opposition. The Spaniards came to the West Indies via the Canary Islands which they began to colonize in 1341, while also seizing the natives, who were treated as inferiors, as slaves. The native Guanches mounted a vigorous resistance to Spanish conquest, and one that led to numerous Spanish casualties, but it was overcome. This difficult conquest has been seen as a conceptual and practical halfway house between the end of the *Reconquista* of Spain and the invasion of the New World.[3]

Vigorous resistance was a major factor in the Americas. However, in the sixteenth and seventeenth centuries, the Spaniards never devoted, nor were in a position to devote, military resources to the New World that in any way compared with their effort in much closer and more urgent European and Mediterranean struggles with France and the Ottoman Turks respectively. While the Spanish conquests of some areas—Cuba (1511–13), central Mexico (1519–21), Peru (1533), and Colombia (1536–9)—were relatively swift, others took far longer. Northward expansion in Mexico was impeded by the Chichimecas in 1550–90 and, further north, by the Spanish discovery that there was no gold-rich civilization to loot. Limited opportunity ensured that the effort put into overcoming native resistance was far less than it might have been. The Spaniards also encountered problems in Central America. Cortés himself led a costly campaign in Honduras in 1524. Guatemala was conquered by 1542. Although much of the Yucatán, the center of Mayan civilization, was overrun in 1527–41, the Itzás people of the central Petén there were not defeated until 1697. In some areas, such as southern Chile, the Spaniards were never successful, ensuring that Native–Spanish relations in Latin America were not simply defined by the control generally exercised by the latter. The first Spanish fort at St. Augustine in Florida, built in 1565, was burned down by the Timucua the following spring.[4]

Similarly, in Brazil, which they "discovered" and claimed in 1500, the Portuguese made only slow progress in extending control in the interior at the expense of the Tupinambá and Tapuya. However, as in Africa, the Portuguese were helped by rivalries between tribes and, indeed, by the alliance of some. Portuguese muskets were of little value against the nomadic Aimoré, mobile warriors who were expert archers and well adapted to forest warfare.[5]

Labor Supply in the Americas

After the initial conquest stage on the mainland in the first half of the sixteenth century, Spanish and Portuguese territorial expansion in the Americas was therefore slower. This had major implications for labor supply. Natives who were willing to supply goods to the Westerners by barter were unprepared to provide continual labor on plantations. As alternative ways of obtaining labor, raiding was more common than the purchase of native slaves, although the latter was significant. The benefits gained from the sale of slaves helped to destabilize Native society by encouraging conflict between tribes in order to seize people for slavery, which was a process also seen in West Africa. At the same time, such conflict between tribes was already well established. The seizure and incorporation of outsiders was an established part of Native society. For example, the Maya of the Yucatán had slaves. Moreover, trading with Westerners was a way to win their support in intra-native warfare. A major change associated with the Westerners, both in the Americas and in Africa, was that captured enemies and other Native slaves were increasingly sold to the Westerners, rather than incorporated into Native households or slaughtered. Sale to Westerners had consequences for population growth, although it is difficult to assess their extent.

Western raiding into unconquered areas in the Americas was widespread and continued for centuries.[6] Raiding, however, was far from easy. Raiders were resisted or fled from. This was a key aspect of the degree to which it was not only in Africa that the slave trade was in part shaped by non-Western responses. There were also security aspects, with those enslaved able to hope they could flee to the areas from which they had been seized. In addition, there was a major rebellion in the leading Brazilian sugar-producing area in 1567. There were other problems affecting reliance on Native slaves and associated slave raiding. Some of the areas into which raids were conducted—for example, the interiors of Brazil, Honduras, and Nicaragua—were distant from the coastal centers of plantation agriculture. Moving them involved losses, and the risk of losses, whether due to flight, accident, or disease. Moreover, the slaves had to be fed on the lengthy journeys. Native slaves were most significant in frontier regions distant from the points of arrival of African slaves—regions such as Amazonia in Brazil and northern Mexico.

Alongside supply problems, control over Native labor within the area of Spanish rule was affected by royal legislation, which both sought to address clerical pressure to treat the Natives as subjects ready for Christianization and not as slaves, and also to assert royal authority by protecting peasants and other subjects against landowners. As a result, following the precedent of the decision, in 1477, by Queen Isabella of Castile to order the freeing of Natives from the Canary Islands who had been sold as slaves in Castile, native slavery was formally abolished in the *Leyes Nuevas* (New Laws) of 1542. The Portuguese government followed with legislation in 1609, and, more definitively, in 1758.[7] The implementation of edicts, such as the Laws of Burgos of 1512 which regulated labor demands, however, took time and was frequently ignored by local officials and landowners. This was always a key element in the treatment of labor: the unwillingness to provide rights to which workers were entitled. Moreover, in Spanish America, systems of tied labor, especially the *encomienda* (land and Native families allocated to colonists) and of forced migration, notably the *repartimiento*, under which a part of the male population had to work away from home, represented de facto slavery. They certainly led to a caste system based on ethnicity and the different histories they gave rise to. This caste system has played a key role in Latin American history, one which has proved very important in its recent politics, not least, but not only, in elections in Bolivia, Peru, and Mexico.[8]

The difficulty of ensuring sufficient numbers of malleable workers encouraged the import of labor from the Old World. There was more than one source. Portuguese peasants were moved to Brazil, but were not available in sufficient numbers nor for slave labor. Initially, part of the labor needs were supplied instead by Moors captured in the conflict with Portugal and Spain, which was frequent in the fifteenth and sixteenth centuries, in southern Spain, North Africa, and the Mediterranean. Although a demonstration of the diversity of slavery and the range of movements this could give rise to, the capture of Moorish slaves, however, proved only a limited source, not least due to Portuguese and Spanish failure in North Africa in the sixteenth century. The Spaniards faced increased and successful opposition there from the late 1520s, being driven from Bougie and from the Peñón d'Argel position dominating Algiers in 1529, while Portuguese expansion in Morocco ended in total and crushing defeat in 1578.[9]

African Slaves

Reliance was eventually placed by both the Portuguese and the Spaniards on African slaves who, therefore, were important to the colonization of the Americas. Between 1500 and 1800, and in stark contrast to the situation thereafter and today, close to four-fifths of the immigrants to the Americas were Africans. The spread of African slavery in the Spanish New World started, and at a modest level, into and in the 1510s. At this stage, this movement was not so much in order to deal with the labor problems created by the death of much of the Native population of the West Indies as, in the 1510s, the Spaniards satisfied such labor needs primarily from other Caribbean islands. This process was extended to the mainland, where there was large-scale slaving among Natives in Honduras and Nicaragua. The former was to satisfy Caribbean demand and the latter to provide slaves for Spanish-ruled Peru, although also to Panama and the Caribbean. This method of labor provision was later extended to North America.[10]

Initially, Africans were shipped into Spanish America via Spain, but, in 1518, *asientos* (licenses) were granted by the Crown for the direct import of slaves from Africa to Spanish America. The supply to Portuguese and Spanish America of African slaves, who had to be purchased in Africa and brought across the Atlantic, was, however, initially more expensive than that of Native slaves. As a result, the Africans were often used not as plantation slaves, but as house slaves, a form of high-value slavery that indicated their cost. As a reminder from the outset of the variety of roles that Africans were to take in the Americas throughout, some fought as conquistadors—soldiers for Spain or Portugal.[11]

Such variety in the use of Africans undercut racial typecasting, but far less so than it should have done. This racial typecasting drew on a Western tradition of presenting the inhabitants of tropical lands as strange creatures, including humans with different characteristics. This tradition rested on Greek accounts of mythical peoples in distant areas, not least Aristotle's claim that it would be too hot for humans to live in the Tropic. Races depicted on medieval Christian maps, such as the Hereford *mappa mundi* (map of the world, 1290s) attributed to Richard of Haldingham, included (in Africa and Asia) the dog-headed Cynocephati, the Martikhora (four-legged beasts with men's heads), the shadow-footed Sciopods, the mouthless Astomi, the Blemmyae (who had faces on their chests), the one-eyed Cyclopes, the Hippopodes (who had horses' hooves), and the cave-dwelling Troglodytes. A one-eyed "Monoculi" was still depicted in West Africa, in Sebastian Münster's *Geographia Universalis* (1540). The very humanity of the inhabitants of these regions was at stake in such depictions. At best, there was a sense of disturbing strangeness, a sense that played a continuing role in the Western response to Africa and Africans. These maps provided ideas about what lay beyond the confines of the known world and about how what was claimed as the latter could be interpreted.

Theologians disagreed about whether Christian redemption would extend to all the "monstrous races." Racial typecasting, indeed, was linked to the treatment of those who were non-Christian. They were regarded as inherently less worthy because the Christian message had not been offered to them or been accepted by them. Indeed, African slavery was widely considered by Western thinkers as justified in natural law. This approach drew on Aristotle's argument that people who were inferior because, due to a defective rational faculty, they could only use their body, were, by their very nature, slaves. Enslavement in this approach, therefore, was not a treatment of equals, but of those who inherently should be slaves—a very useful and totally inaccurate argument.

By the mid-sixteenth century, the demographic and economic situation had changed. Rather than providing a marginal part of the labor force for Spanish America, Africa became steadily more important as a source of slaves, not least because it was believed that Africans were physically stronger than Natives. Nevertheless, African slaves remained more expensive than Native labor, which could be variously controlled, including by service due to debts. By 1570, there were probably only 20,000 African slaves in Spanish America.[12] In comparison with future movements of slaves, fewer than 5 percent of the total "exported" from Africa in the Atlantic trade between 1450 and 1900 were moved prior to 1600. However, the trend in both slave numbers and slaves shipped was upwards, especially from the mid-sixteenth century, with the number of African slaves in Cuba rising from about 1,000 in 1550 to about 5,000 in 1650. The upward trend reflected demand from both Spanish and Portuguese colonies as well as the availability of supply. It is highly regrettable to refer to people as commodities in this fashion, but that was the nature of the slave trade and a way to understand its inhumane character.

Where African slaves were not available, the native population was also treated harshly, whether or not they were referred to as slaves by the Russians. This was true from the 1740s of the Aleuts of the Aleutian Islands. In addition, there was an assault on animal life that could be used such as sea otters, sea cows, sea lions and fur seals in the Aleutians,[13] and fur-bearing animals in continental North America.

Demand

Unlike Native Americans, many African slaves came from cultures that regularly worked with iron implements and in which cattle were raised. These slaves were used for a variety of tasks across Brazil and Spanish America. This reflects the mistake in assuming that the slave trade was primarily in order to produce labor for plantation agriculture, a mistake that draws on a simplification that is unfortunately all too common in discussion of the subject. In some parts of Spanish America, agricultural work was indeed important. This was usually so for cash crops, such as sugar in the valleys of northern Peru and wine on the Peruvian coast. Profits from such crops, which had been introduced by the Spaniards as part of their colonization from the sixteenth century, covered the cost of purchasing and then transporting slaves. In Peru, Africans were judged more suitable for the heat and humidity of the coastal valleys than Native Americans, who were generally used in the higher and drier terrain of the Andean chain with which they were more familiar.

As a result of such choices, the slave trade was part of the process by which Westerners reconceptualized and responded to the geography and demographics of the areas they conquered. The increase of knowledge thus became an aspect of the use of labor, with the control offered by slavery helping to ease the response. This was one way in which the wider process later referred to as the Scientific Revolution was linked to the extension of slavery. The mass of information received from transoceanic voyages, and the need to order and employ it, encouraged a rethinking and reusing of the world by Westerners. This rethinking involved changes in navigation and mapping, notably the development of the Mercator projection. There was also much information produced by discoveries on land.

Indeed, the extent to which the New World challenged existing ideas has led to the claim that the Scientific Revolution began with the Spanish response to their new lands. The Spanish Crown's interest in profiting from its new territories encouraged the exploration of nature there, being part of the process by which far-reaching state systems extended human interactions with the environment.[14] Slaves provided a key means for this interaction.

Slaves were used to produce food to be sold to the cities and to the mine towns, such as the great silver-producing center of Potosi in modern Bolivia, which was a crucial source of bullion for the Spanish Empire and for the

Western trading system. Silver from the New World financed the wars of the Spanish state and enabled Europe to fund trade with areas such as China and India with which they had a negative balance of payments. Mine towns were an area of work for slaves, although the prime labor supply there was from Native Americans. Africans, reflecting their higher value, tended to be used in these mine towns for refining and as overseers. In the cities, slaves were employed for a variety of tasks, including as craftsmen, servants, and laborers.[15]

Black people remained a smaller group than Native Americans across most of Spanish America. In central Mexico, the percentage of the population who were Native Americans in 1646 was 87.2, while Spaniards, whether born in Mexico or immigrants, amounted to 8 percent, *mestizos* (mixed) were 1.1 percent and *pardos* (wholly or partly black) 3.7 percent. In the mid-1740s, the respective percentages were 74 and then about 9 each, while as far as the whole of Mexico was concerned, the population in 1810 has been estimated at about 6,121,000: 3,676,000 Native Americans, 1,107,000 Spaniards, 704,000 *mestizo*, and 634,000 *pardos*. Thus, those who had come from Africa, or their descendants, were only a minority of the Mexican population.[16] However, where disease had really ravaged the Native population—for example, on Puerto Rico and Hispaniola—the black population by 1600 was greater than that of the surviving Taînos.

In the Portuguese colony of Brazil, which was largely the coastal regions of modern Brazil south of the Amazon, the initial emphasis was also on the use of Native labor, just as, in the Canaries, the Guanches were initially used by the Spaniards as slaves on the sugar plantations. Enslaved Native labor was in part used for producing food to feed towns such as São Paulo, with the Portuguese exploiting divisions between and within Native groups in order to obtain the slaves.[17] However, alongside rising labor demands for mining and agriculture, as sugar cane was introduced in Brazil and became more important, it proved necessary to supplement slave raiding into the interior with the import of slaves from West Africa and Angola. The Portuguese had bases in each of these—notably Accra, Axim, Elmina, Fort São Sebastião (Shama) in West Africa, and Luanda in Angola. Aside from exporting slaves, the Portuguese used others in their plantations in Angola. The Native slave population of Portuguese Brazil was hit hard by a smallpox epidemic in 1560–3 and a measles epidemic in 1563. As a result, needing labor for the expanding sugar economy, Portuguese Brazil in the last quarter of the century imported about 40,000 African slaves.[18]

Supply

It is instructive to consider demand before supply because that explains why Westerners created and responded to opportunities in Africa, with the creation and the response frequently proving to be closely linked, if not being two sides of the same coin. The changing nature of the opportunities offered by Africa is instructive. Portuguese expeditions along Africa's Atlantic coast in the fifteenth century had been motivated primarily by a search for gold and not for slaves. These expeditions gathered pace in part thanks to the support of the influential Prince Henry of Portugal (1394–1460), who in 1416 founded what was to be an important school of navigation. In 1434, Gil Eanes became the first European to navigate successfully round Cape Bojador, which had long been regarded as an obstacle to navigation.

Gold was particularly important in a metallic money system, a system in which the value of money was measured by its metallic content, rather than by the paper value that is the focus of modern attention. Gold was the most valued metal. Among the illustrations on the African portion of the *Catalan Atlas* of 1375 was a depiction of the fabled King of Mali, Mansa Munsa. The text read, "So abundant is the gold in this country that the lord is the richest and noblest king in all the land." Legend had it that Mali possessed dazzling quantities of gold. A map of 1413 by Mecia de Viladestes reflected the expansion of knowledge of West Africa as Western explorers moved south along the African coast. It also depicted hope in the shape of the "river of gold" that would apparently provide a trading route from the Atlantic into the West African interior.

This map indicates interest in obtaining gold and other goods without the intermediary of North African Muslims. Expansion along the West African coast, therefore, was an aspect of geopolitics, specifically the conflict between Christendom and Islam. In practice, it did not prove easy for the Portuguese to gain direct access to the gold, which, instead, was exchanged for salt from the Saharan salt mines that was brought by Moorish traders using camels, some of which were then eaten. The Portuguese, nevertheless, also gained entry into the gold trade of West Africa. Obtained both from the River Gambia and on the Gold Coast, gold acted as a major spur for the slave trade, because slaves acquired elsewhere on the West African coast became useful in order to have something to sell so as to acquire gold. The gold trade of the base of São Jorge da Mina (Elmina) on the Gold Coast, founded in 1482, was controlled as a monopoly of the Crown of Portugal but with the Portuguese paying regular tribute to the Caramança of Mina.[19]

The Portuguese first met black Africans in Senegambia, the region from the mouth of the River Senegal, which the Portuguese reached in 1445, to that of the River Gambia. It was from there that the first black slaves directly acquired by the Portuguese came. They obtained them by raiding but, more commonly, by purchase as they sought to benefit through adaptation to the local power system.[20]

The sale of slaves brought from Africa either to Portugal or to its Atlantic Ocean colony of Madeira became important, Africans feeding into the pattern already established for captured Moors. The Portuguese exported significant numbers of slaves from Africa from the 1440s, with between 140,000 and 170,000 slaves imported into Portugal by 1505. Most worked in domestic service in Portugal or on the sugar plantations of Madeira. From Lisbon, West African slaves were reexported to Spanish and Mediterranean slave markets, notably those of Valencia and Majorca.

This use of slaves from West Africa for domestic service amplified the already existing purchase in the slave markets on the North African coast, such as Tunis and Tripoli, of slaves brought across the Sahara. It is instructive to note the frequency with which black slaves occur in the paintings of Paolo Veronese (1528–88), paintings that testify to the magnificence of Renaissance Venice. Black house slaves are featured in *The Supper at Emmaus* (*c*.1575–80) and *Judith and Holofernes* (*c*.1580–5). In *The Adoration of the Kings* (1573), one of the kings as well as a servant are black. The executioner in Veronese's *The Martyrdom of Saint George* (*c*.1565) has a black slave.

In time, the Portuguese–West African gold and slave economy was to be transformed into a slave economy focused on Brazil, which was formally claimed by Portugal in 1500. The emphasis on slaves reflected labor needs, especially the demands of sugar production which ensured that Brazil came rapidly to play a key role in the transatlantic slave trade. Sugar production had been developed within the Western world in the Atlantic islands: Madeira, the Azores, the Cape Verde Islands, and the Canaries. The Portuguese settlement of Madeira began in 1424, of the Azores in the 1427, and of the Cape Verde Islands in the late 1450s. On the island of São Tomé, which Portugal settled in 1471, large-scale sugar plantations were established, benefiting from slaves brought from nearby Africa.

It was from these islands that the plantation system was transferred to Hispaniola in 1503, and to Brazil in the 1530s.[21] Brazil, where the Portuguese had initially focused on cutting down tropical hardwoods for export, rapidly supplanted Madeira as the leading producer of sugar in the Portuguese world, employing a comparative advantage due to the ready availability of slave labor,[22] as well as relatively fresh soil in plentiful quantities. Madeira and São Tomé were restricted sites, unlike Brazil. In a pattern that was also to be seen in the spread of Western agriculture in the nineteenth century, new labor and new soil that could be readily used were key elements in enabling Westerners to profit from areas that were newly gained. These factors enabled Portugal to benefit from its new colony: the number of sugar mills in Brazil rose from 60 in 1570 to 192 in around 1600.

Northeast Brazil was the center of sugar production: there was a lengthy harvest season, relatively mild weather and soils not already denuded by much cultivation. North-east Brazil was also close to West Africa, a reminder of the extent to which oceans could link rather than separate. It was easier to cross the Atlantic than the African continent. As a result, continents (i.e. landmasses), as a way to define space, are an artificial construct.[23] Indeed, as a consequence of the development of links in the Atlantic economy, West Africa was to become closer in some respects to Brazil than it was, for example, to most parts of Africa, let alone distant South-East and East Africa. Relatively short slaving voyages across the Atlantic were particularly valuable because they reduced the need for credit in bridging the period between the purchase and sale of slaves. As a result, capital requirements for the trade were lower than they would otherwise have been. Death rates among the slaves were also generally lower on shorter voyages, and this decreased the cost and increased the profitability of the trade. In the 1570s, the percentage of Africans among the slaves in north-east Brazil increased, such that, by the mid-1580s, about one-third of the slaves were Africans. The latter became the majority of the slaves by 1620 and the slave population of Brazil rose to about 150,000 in 1680. The slave trade provided royal revenues to the crown of Portugal: aside from slaves moved on the royal account, private slave traders were taxed.

In both Brazil and Spanish America, the high death rate and low fertility rate of African slaves once arrived reflected the cruel hardships and disorientation of enslavement, transport, and labor. These rates ensured that it was necessary, in order to sustain the numbers needed for work, to import fresh slaves. This continual import of slaves affected the nature of slave society, sustaining its African character—for example, the traditional pattern of marriage customs, religious beliefs, and related ideas about kin and family, and thus its foreignness to native societies.[24]

The Atlantic slave trade from Africa was a new variant on the longstanding pattern of slavery and slave trade within Africa; a pattern, however, that is relatively obscure, and certainly far more so than the Atlantic slave trade. There is a need to emphasize the term "variant" because there is room to suggest that the possibilities of overseas trade encouraged new ways to exploit existing patterns of social stratification and dependency, as well as raiding and conflict in order to obtain slaves. At the same time, slave owning and hierarchical monarchies did not originate at the time of the Atlantic slave trade, but were the case in parts of Africa—for example, the western Sudan, from the eleventh century, if not earlier. This point, however, might suggest not indigenous (local) origins for sub-Saharan slavery, but rather the impact of the trans-Saharan slave trade from North Africa—in short, of outsiders. In this case, however, they were Muslims, rather than Christians operating along the Atlantic coast. Muslim influence spread south across the Sahara and into the *sahel* in the tenth and eleventh centuries. Neither conflict nor slavery was dependent on the Western presence, and this remained the case.

For the sixteenth century, this was true, for example, of the destruction in West Africa of Mali by the Songhai Empire in 1546 and of the subsequent overthrow of the latter by the Moroccans in 1591. It was also the case in the

Horn of Africa (Ethiopia and Somalia) of the destructive attacks on Adal and Ethiopia by the pagan Galla from the Ogaden. The Moroccan dimension to enslavement was given renewed energy from the late sixteenth century. Having defeated the Portuguese in 1578, the Saadian dynasty expanded not as an Atlantic power, but as one that brought renewed energy to the trans-Saharan slave trade. The capture of the Tuat and Gourara oases of modern Algeria in the mid-1580s expanded control over trans-Saharan trade, while the Songhai empire of the Niger Valley was overthrown in 1590-1 and Moroccan power established there.

Separately, in 1619–68, an independent privateering republic was proclaimed at Salé in Morocco by Moriscos who had fled from Spanish Morocco. Their fleet, the Salé Rovers, journeyed far in search of slaves and loot, including to Iceland.

It is unclear how far labor shortages in Africa encouraged enslavement in order to secure labor (a system analogous to the "Second Serfdom" in Eastern Europe), and how far this system then helped provide slaves to Western traders. The issue is related to the extent to which slaves and, more generally, people were a form of wealth, both a source and a symbol of wealth in sub-Saharan Africa. In part, this was because people were scarcer than land, which only took on value when it was farmed or when valuable minerals and metals were found on it. People were aligned in kinship groups; and those from other groups were acquired to enhance power and value. Furthermore, as a result, Africans could readily commodify slaves for use for barter and as money.

Not only were economic factors at issue. The majority of Africans who were sold as slaves were captured in warfare, but some were also enslaved as a result of judicial punishments. Those who were captured had thus lost their tribal identity, and this loss was an important aspect of slavehood, its disorientating character, and cutting people off from their origins.[25] The situation was similar among some other societies—for example, the Maori of New Zealand. The loss of tribal identity therefore occurred before the African slaves were shipped across the Atlantic. However, this loss of slaves' identity was made more abrupt, more disorientating, and more permanent by their being shipped.

On the west coast of Africa, the Europeans obtained slaves largely by trading. The business was initially conducted from on board the slave ships which were anchored in estuaries such as that of the River Gambia or on the coast. The Westerners were not powerful enough to seize slaves by the large-scale raids employed in Central America and Brazil. The Portuguese at first favored such raids in Africa, but abandoned them in the late 1440s because the rulers of Upper Guinea south of the River Senegal were too strong.[26] The Portuguese, indeed, faced a variety of serious problems in Africa, problems that qualify the sense of Western military superiority, a sense based on the misleading idea that firearms swept all before them. African coastal vessels, powered by paddles and carrying archers and javelinmen, which provided missile power of their own, were able to challenge Portuguese raiders on the West African coast. Although it was difficult for them to storm the larger, deeper draught and high-sided Portuguese ships, they were, nevertheless, too fast and too small to present easy targets for the Portuguese cannon. In 1535, for example, the Portuguese were once more repelled when they tried to conquer the Bissagos Islands off the West African coast, and thus to gain a possibly more secure base.

Furthermore, disease was as debilitating for the Westerners in Africa as it was for their native opponents in the New World. About 60 percent of the Portuguese soldiers who served in Angola in 1575–90 died of disease. Most of the rest were killed or deserted. In addition, horses could not survive, which compromised the possibility of the Europeans employing cavalry or dragoons in order to fight or to hunt for natives. European death rates were not only a problem in Angola. In West Africa, the average annual death rate of the factors of the British Royal African Company between 1684 and 1732 was 270 per 1,000.[27]

Moreover, prevailing wind and ocean conditions limited Portuguese access to the African coast south of the Gulf of Guinea, while the extensive coastal lagoons and swamps of West Africa made approaching the coastline difficult, and notably so for deep-draught ocean-going ships. Penetration into the African interior was variously hindered by tropical rainforest in West Africa and, on the coasts of modern Mauritania and Namibia, by pronounced desert conditions.

More generally, the environment was far more difficult for Western conquest than in much of the New World. Whereas Mexico and Peru were populous and had a well-developed agricultural system that could provide plenty of resources for an invader, Africa lacked comparable storehouses, food for plunder, and roads. Mexico and Peru were also more centralized politically, and thus easier for Spain to take over once the ruler had been seized. In contrast, Africa was more segmented and new chiefs could emerge. This helped encourage the high rate of conflict within the continent that fed the slave trade.

Instead of conquest, trade was a more successful means of Western access to Africa and also helped to finance further expansion. It is important not simply to focus on what the Europeans wanted but also on what the Africans were prepared to deliver. The weakness that Europeans had in direct capture meant that they had to buy slaves. Once they gave up capturing a few hundred people at the cost of quite a few casualties of their own, they turned to purchase, and greatly increased the number of the enslaved that they were able to ship. Thus, although it has been argued that the Europeans somehow imposed the slave trade on Africa, there was a considerable degree of African initiative and control. This reflected the degree to which the forced move-ment of people from one place and concentrating them elsewhere was a longstanding practice, one that led visitors to comment on the number of slaves. However, the slavery in Africa might be "milder" because it did not imply any particular sort of labor regime, mostly indeed being little different from the labor of the free, and not plantation agriculture, even if often the surplus from the slaves in Africa might go to private persons, with whom indeed bonds of loyalty could develop, rather than the state.[28]

The profits from the Portuguese trading base of São Jorge da Mina (Elmina) in West Africa, founded in 1482, financed later voyages down the coast of West Africa, such as those of Diogo Cão and Bartolomeu Dias. In 1483, Diogo Cão became the first Westerner to set foot in the kingdom of Kongo. Peaceful relations were established and in 1491 the king was baptized as João I. A syncretic (mixed) blend of Christianity and local religious ele-ments spread rapidly. This syncretism is a good model for understanding much of the more general process of Western impact in Africa. However, conversely, Christian conversion on this scale did not occur elsewhere, despite attempts in Sierra Leone in the early seventeenth century. In 1485, Diogo Cão pressed on to reach the coast of modern Namibia. Dias rounded the Cape of Good Hope into the Indian Ocean in 1488. Discovery of the mari-time pathway to India encouraged interest in the South Atlantic en route.

The possibilities of the slave trade had led in 1486 to the establishment of the *Casa dos Escravos de Lisboa* (Lisbon Slave House). Mina itself was a logistical achievement, prefabricated with stores, timbers, and tiles all pre-pared in Portugal. Later Portuguese bases in West Africa included Axim (1495), Accra (1515), and Shama (1526), and, off the coast, Fernando Poo (1483).[29] The Portuguese led in the trade to and from the sub-Saharan coasts of Africa, not least because the Spaniards ceased to be traders there in 1479 under the Treaty of Alcáçovas. In part, this reflected the definition of Portuguese and Spanish zones of influence, a definition that was to be important to the ability of the two monarchies to cooperate. This definition was carried forward by the Treaty of Tordesillas in 1494 under which Pope Alexander VI divided the western hemisphere between Spain and Portugal, although the difficulty of fixing latitude created issues of interpretation, while France did not accept the division. America, with the exception of Brazil, was allocated to Spain. In 1529, the Treaty of Saragosa followed up by dividing the eastern hemisphere between the two powers.

Once the Spaniards had established themselves in the Americas, this Spanish absence from the West African coast could be maintained because of the longstanding Spanish willingness to purchase slaves from others. Thus, licenses were given to the Portuguese to import slaves into the Spanish New World, and the nature of labor requirements helped ensure a preponderance of male slaves despite interest in ensuring a gender balance.[30] Moreover, the

Spaniards were able to secure slaves in the Americas, including from the children of slaves and from raiding among the Native American population beyond the bounds of Spanish power. The coerced labor of the subject native population was also important.

The English Enter the Picture

The Portuguese position in West Africa was challenged not by the Spaniards, but by the English in the mid-sixteenth century, from the reign of the Catholic Mary I (r. 1553–8), wife of Philip II of Spain (r. 1556–98). The pioneers of English slavery were merchants based in lower Andalusia in Spain at the end of the fifteenth and beginning of the sixteenth centuries, with William de la Founte in 1490 proving the earliest documented English slaveholder there. At a very different level, Henry VII (r. 1485–1509) kept some black domestic servants. The first recorded English colonist in Hispaniola was Nicholas Arnold in 1508.[31]

The English subsequently made an attempt to break into Portugal's trade with West Africa, and into the profitable slave trade between there and the Spanish New World. John Hawkins, who obtained his slaves in West Africa by raiding rather than purchase (losing men in the process to poison arrows and other means), sold slaves to the Spaniards. However, in 1568, on his third slaving voyage, at San Juan de Ulúa near Vera Cruz in modern Mexico, the presence of the Viceroy of New Spain led to a Spanish attack on what was, in the official view, an unwelcome interloper. This attack helped ensure that the venture made a large loss. Only two English ships survived the attack.

The unwillingness of Elizabeth I of England (r. 1558–1603) in the early years of her reign to confront directly the imperial interests of Portugal and Spain encouraged a reliance by her subjects on unofficial or semi-official action such as privateering. England did not go to war with Spain until 1585. By then, Portugal's position had been transformed. In the aftermath of the total Moroccan defeat of the invading Portuguese army at Alcazaarquivir on August 5, 1578 and the death there of its commander, the reckless and childless Sebastian I, Philip II of Spain also became Philip I of Portugal in 1580. He enforced his claim to the succession with the conquest of Portugal in one of the most rapid and decisive military campaigns of the century, one that was to transform the geopolitics of the slave trade.

This dynastic link between Spain and Portugal continued until a successful rebellion in Portugal in 1640. As a result of this link, the supply to Spanish America of African slaves obtained by the Portuguese was facilitated. However, it was also possible for those at war with Spain, such as England between 1585 and 1604, and 1624 and 1630, and, more significantly, the Dutch from 1565 to 1609 and 1621 to 1648, to breach the Portuguese monopoly in the slave trade without fear of admonition from their home governments.

In the late sixteenth century, the English commitment to the slave trade was far less than it was to be in the seventeenth century. Most English voyages in this period to West Africa were for pepper, hides, wax, ivory, and in search of gold—goods to be shipped to England—and not for slaves, a trade that entailed crossing the Atlantic. No English fort was built in West Africa in this period. Indeed, English trade with West Africa did not focus on the slave trade until the mid-seventeenth century.[32] This serves as a reminder of the need to locate the slave trade in the broader pattern of Western expansion and of very different opportunities in the Atlantic economy, as well as those that overlapped.

The Dutch, initially, also played only a modest role in the trade with West Africa. They were subjects of the Emperor Charles V (Charles I of Spain) and then of his son, Philip II of Spain, and therefore affected by Spanish rights and claims. Moreover, the Low Countries were in the forefront of the struggle between Charles (and then Philip) and the rulers of France in the 1540s and 1550s. After they rebelled against Philip in the mid-1560s, the Dutch were primarily concerned with the war for their independence and the naval struggle in home waters. However, in 1594, Philip banned Dutch trade with Lisbon, which encouraged the Dutch to look further afield for profit as well as to exploit the commercial opportunities offered by the Portuguese empire.

France from the 1560s to 1590s was consumed by civil war, the bitter French Wars of Religion. These wars did not prevent French attempts to establish themselves in the New World, but they were unsuccessful in this period. Although the French burnt Havana in 1552, their attempts to establish themselves in Florida were defeated by the Spaniards. French attempts to found bases in Brazil—in 1555 on the site of modern Rio de Janeiro and in 1558 and 1612 at São Luis in northern Brazil—proved short-lived, although the longstanding French attempt to challenge the northern border of Brazil led to the later establishment of a French colony at Cayenne.

The ability of the Spanish and Portuguese to see off repeated attack proved impressive. In the Caribbean, Spanish defensive measures improved in the sixteenth century. As a result, in 1595–6, Sir Francis Drake's last Caribbean expedition—a major attempt at English expansionism—was a failure. Had Spain and Portugal been less successful, then English primacy in the Atlantic slave trade might have started far earlier.

Ships and Maps

The greater experience in long-distance, deep-sea voyaging, and commerce gained in the sixteenth century was an important background to the later expansion of the slave trade and helped ensure the success of the latter. This experience rested on an important improvement in the capability of shipping, which gave the Westerners a powerful comparative advantage over non-Westerners. Late fourteenth- and fifteenth-century developments in ship construction and navigation included the fusion of Atlantic and Mediterranean techniques of hull construction and lateen- and square-rigging, the spread of the sternpost rudder and advances in location finding at sea. Carvel building (the edge-forming of hull planks over frames), which spread from the Mediterranean to the Atlantic from the late fifteenth century, replaced the clinker system of shipbuilding using overlapping planks, contributing significantly to the development of stronger and larger ship hulls which were necessary for successful trade across the Atlantic. The increase in the number of masts on large ships expanded the range of choices for rigging and also provided a crucial margin of safety in the event of damage to one mast. Developments in rigging, including an increase in the number of sails per mast and in the variety of sail shapes, permitted greater speed, a better ability to sail close to the wind, and improved maneuverability.[33]

Navigational expertise increased, moreover—a reminder of the long-standing relationship between information, knowledge, and the slave trade. The greater range of Western trade made such an increase in expertise necessary. Thanks to the use of the magnetic compass, as well as the spread from the Mediterranean to Atlantic Europe of astrolabes, cross-staffs, and quadrants (which made it possible to assess the angle in the sky of heavenly bodies) and other developments in navigation, such as the solution in 1484 to the problem of measuring latitude south of the Equator, it was possible to chart the ocean and to assemble knowledge about it. This knowledge was an important prelude to the further development of the slave trade, not least because better charts helped reduce the risk of voyaging, and thus the hazards of sailing.[34] In 1516, the explorer Amerigo Vespucci's nephew, Juan, was instructed by the Spanish government to produce a *pardón real* (official royal chart), a work that was frequently updated to take note of new reports from navigators. The mapping of the New World provided further point to that of the oceans. Columbus's pilot on his second voyage, Juan de la Cosa, is usually held to have produced the first map to show the discoveries, but it may have appeared later than the traditional date, 1500. In 1502, the Cantino map depicted the Americas and also West Africa, the coast of which was revealed in greater detail.

Although in Atlantic Africa the Portuguese use of native interpreters gave them access to African views about their own societies,[35] the interior of Africa was largely unknown to Westerners. Giacomo Gastaldi's eight-sheet map of Africa, published in Venice in 1564, the largest map of Africa yet to appear, reflected journeys into the interior—for example, those of the Portuguese into Ethiopia, but these were patchy and the map also depicted nonexistent large lakes. Some information came to Europe from the Islamic world, where there was a long tradition of using captives and renegades.

The relatively recently developed process of printing spread the resulting information. The *Descrittione dell' Africa* (Venice, 1550) by the Arab scholar known as Leo Africanus was an important source for Western mapping, but his errors included the idea that the River Niger flowed westwards, a belief linked to the conviction that a large lake must be its source.[36]

There was also the issue of how information was depicted and used. In 1569, Gerard Mercator, a Flemish mathematician, produced a projection that treated the world as a cylinder, so that the meridians were parallel, rather than converging on the Poles. Taking into account the curvature of the Earth's surface, Mercator's projection kept angles, and thus bearings, accurate in every part of the map. A straight line of constant bearing could thus be charted across the flat surface of the map, a crucial tool for navigation. This projection made most sense for compass work, pilotage, and navigation, especially in mid-latitudes, which included those from West Africa to the West Indies. As the shape of the world was increasingly grasped, so the opportunities for profit appeared more realizable and immediate. The growth in transatlantic trade, in goods and slaves, was to reflect this. As the trade became more regular and predictable, so it became more dependable as a source of labor and profit.

Conclusions: Race and Trade

In turn, this growth in the slave trade fortified the perception and treatment of sub-Saharan Africans as slaves. Traditional ideas were employed to help shape the developing situation. Rationality was seen as the main distinction between humans and animals. Moreover, the soul was presented as moral as a result of the reason given to humans by God. Thus, the ability to process information and acquire knowledge was an aspect of the divine plan, as well as a means to distinguish among creation. The capacity for human fall from divine grace was already present from the Biblical story and slavery was condoned in the Bible. This capacity was offered anew by human behavior that was apparently, in hostile eyes, similar morally to that of animals. This categorization was then applied to non-Western native peoples who were presented as irrational.[37]

Overlapping with this approach, but more specific in its application, racial categorization was deployed as a way to cope with what was now, in the shape of slaves and skin color, a more prominent "otherness." "Blackness" had proved a somewhat slippery concept for Westerners, who tended to see some of their own number as dark-skinned. As a result of the slave trade, however, black Africans were stereotyped, and many African cultural practices were misunderstood and recast in a negative light. As slaves became more common in the Western world, so there was a hardening of attitudes to slavery. Denigration of Africans as inferior and uncivilized was related to their pigeonholing in occupations linked to physical prowess and thus to slavery.[38] Whiteness was fixed as both origin and norm in the Western theories of blackness expressed in this period.

The same process took place in the Islamic world, with discrimination there directed against darker Arabs as well as Africans. This discrimination has lasted to the present day, with Omanis frequently referred to by Gulf Arabs in terms reflecting the belief that they are at least part African. The Sultanate of Oman was a key player in the slave trade from East Africa to the Islamic world.[39] In Arab towns, especially small towns, that were major centers of slave trading—for example, the Tunisian town of Kebili—there is sometimes a high percentage of inhabitants who are black or of mixed background.

The relationship between racist attitudes and the grasping of economic opportunity is a complex one and causes probably operated in both directions. Separately, drawing attention to the widespread nature of unequal, and frequently coerced, labor relationships in this period does not, and should not, serve to downplay the horrors of slavery, but instead to contextualize them. Slavery was a key aspect of coerced labor, but not the only one. As a consequence, the slave trade was both distinctive and a part of a wider movement of coerced labor. The conditions of the slave trade might well appear the harshest, and certainly were compared to the movement of indentured servants. However, there is no comparable contrast in the case of transported convicts. The treatment of the latter—compulsory work—opens up anew functional questions of the definition of slavery. In terms of the Atlantic slave trade, there is the geographical distinction between the labor flows from Europe and those from Africa. The former, generally, did not include slaves, although there were exceptions, notably Moors and sub-Saharan Africans moved through Western entrepôts. In contrast, those who were taken from Africa to the New World were overwhelmingly slaves. They are the subject of this book.

Notes

1 J.D. Daniels, "The Indian Population of North America in 1492," *William and Mary Quarterly*, 49 (1992), 298–320.

2 K. Sale, *The Conquest of Paradise* (London, 1990); M. Livi-Bacci, "Return to Hispaniola: Reassessing a Demographic Catastrophe," *Hispanic American Historical Review*, 83 (2003), 3–51; S.A. Alchon, *A Pest in the Land: New World Epidemics in a Global Perspective* (Albuquerque, NM, 2003).

3 F. Fernández-Armesto, *Before Columbus: Exploration and Colonization from the Mediterranean to the Atlantic, 1229–1492* (Philadelphia, PA, 1987), 212–13.

4 "Native" when capitalized from hereon refers to Native Americans, descended from the pre-Iberian conquest population and not to those born in the Americas, who might include the descendants of the conquerors.

5 J. Hemming, *Red Gold: The Conquest of the Brazilian Indians, 1500–1760* (2nd ed., London, 1995), 72–3, 78–9, 90–3.

6 J.M. Monteiro, "From Indian to Slave: Forced Native Labour and Colonial Society in Sao Paulo During the Seventeenth Century," *Slavery and Abolition*, 9 (1988), 105–27.

7 R.A. Williams, *The American Indian in Western Legal Thought: The Discourses of Conquest* (New York, 1990).

8 O.N. Bolland, "Colonization and Slavery in Central America," in *Unfree Labour in the Development of the Atlantic World*, ed. P.E. Lovejoy and N. Rogers (Ilford, 1994), 11–25.

9 D. Wheat, "Mediterranean Slavery, New World Transformations: Galley Slaves in the Spanish Caribbean, 1578–1635," *Slavery and Abolition*, 31 (2010), 327–44.

10 B. Rushforth, "'A Little Flesh We Offer You': The Origins of Indian Slavery in New France," *William and Mary Quarterly*, 60 (2003), 777–808.

11 M. Restall, "Black Conquistadors: Armed Africans in Early Spanish America," *The Americas*, 57 (2000), 171–206.

12 C.A. Palmer, *Slaves of the White God: Blacks in Mexico, 1570–1650* (Cambridge, MA, 1976), 28.

13 R.T. Jones, *Empire of Extinction: Russians and the North Pacific's Strange Beasts of the Sea, 1741–1867* (Oxford, 2014).

14 A. Barrera-Osorio, *Experiencing Nature: The Spanish-American Empire and the Early Scientific Revolution* (Austin, TX, 2006), 2; J.F. Richards, *The Unending Frontier: An Environmental History of the Early Modern World* (Berkeley, CA, 2003).

15 F. Bowser, *The African Slave in Colonial Peru, 1524–1650* (Stanford, CA, 1974). More generally, M.J. MacLeod, *Spanish Central America: A Socioeconomic History, 1520–1720* (Berkeley, CA, 1973).

16 P. Bakewell, "Spanish America: Empire and its Outcome," in *The Hispanic World*, ed. J.H. Elliott (London, 1991), 74–5.

17 J. Monteiro, *Blacks of the Land: Indian Slavery, Settler Society, and the Portuguese Colonial Enterprise in South America* Cambridge, 2018).

18 S.B. Schwartz, "Indian Labor and New World Plantations: European Demands and Indian Responses in Northeastern Brazil," *American Historical Review*, 81 (1978), 72–3; A. Marchant, *From Barter to Slavery: The Economic Relations of Portuguese and Indians in the Settlement of Brazil, 1500–1580* (Baltimore, MD, 1942).

19 J. Vogt, *Portuguese Rule on the Gold Coast, 1469–1682* (Athens, GA, 1979).

20 H.L. Bennett, *African Kings and Black Slaves: Sovereignty and Dispossession in the Early Modern Atlantic* (Philadelphia, Penn., 2019).

21 S.M. Greenfield, "Madeira and the Beginning of New World Sugar Cane Cultivation and Plantation Slavery," in *Comparative Perspectives on Slavery in New World Plantation Societies*, ed. V.D. Rubin and A. Tuden (New York, 1977), 536–52.

22 L. Felipe de Alencastro, *The Trade in the Living: The Formation of Brazil in the South Atlantic, Sixteenth to Seventeenth Centuries* (New York, 2019).

23 M.W. Lewis and K.E. Wigen, *The Myth of Continents: A Critique of Metageography* (Berkeley, CA, 1997).

24 C.A. Palmer, "From Africa to the Americas: Ethnicity in the Early Black Communities of the Americas," *Journal of World History*, 6 (1995), 223–36, esp. 235–6.

25 T. Green, *The Rise of the Trans-Atlantic Slave Trade in Western Africa, 1300–1589* (New York, 2012).

26 M. Newitt, *A History of Portuguese Overseas Expansion, 1400–1668* (Abingdon, UK, 2005), 45.
27 K.G. Davies, "The Living and the Dead: White Mortality in West Africa, 1684–1732," in *Race and Slavery in the Western Hemisphere: Qualitative Studies*, ed. S.L. Engerman and E.D. Genovese (Princeton, NJ, 1975), 88–93.
28 J.K. Thornton, *A History of West Central Africa to 1850* (Cambridge, 2020).
29 A.W. Lawrence, *Trade Castles and Forts of West Africa* (London, 1963).
30 D. Wheat, "*Nharas* and *Morenas Horras*: A Luso-African Model for the Social History of the Spanish Caribbean, *c.* 1570–1640," *Journal of Early Modern History*, 14 (2010), 129–30.
31 Ex. Inf. Gustav Ungerer.
32 K.R. Andrews, N.P. Canny, and P.E.H. Hair, eds., *The Westward Enterprise: English Activities in Ireland, the Atlantic and America, 1460–1650* (Liverpool, 1978).
33 I. Friel, *The Good Ship: Ships, Shipbuilding and Technology in England, 1200–1520* (London, 1995).
34 On these risks, P.E. Pérez-Mallaína, *Spain's Men of the Sea: Daily Life on the Indies Fleets in the Sixteenth Century* (Baltimore, MD, 1998); C.A. Fury, *Tides in the Affairs of Men: The Social History of Elizabethan Seamen, 1580–1603* (Westport, CT, 2002).
35 P. Russell, "*Veni, Vidi, Vici*: Some Fifteenth-Century Eyewitness Accounts of Travel in the African Atlantic before 1492," *Historical Research*, 66 (1993), 115–28.
36 N.Z. Davis, *Trickster Travels: A Sixteenth-Century Muslim between Worlds* (New York, 2006).
37 E. Fudge, *Brutal Reasoning: Animals, Rationality and Humanity in Early Modern England* (Ithaca, NY, 2006).
38 T.F. Earle and K.J.P. Lowe, eds., *Black Africans in Renaissance Europe* (New York, 2005).
39 A. Sheriff, *Dhow Cultures of the Indian Ocean: Cosmopolitanism, Commerce and Islam* (London, 2010).

3 The Slave Trade Expands Greatly

Plantation Crops

The major growth in the Atlantic slave trade in the seventeenth century was to be driven by the expansion of New World exports to Europe, and it is appropriate to begin with these economic forces, rather than with the process of Western colonization itself. The opportunities, pressures, equations, and profits of demand played key roles in the growth of the slave trade, more particularly by establishing close and profitable linkages between different parts of the Atlantic world. The particular labor demands of individual crops and trades ensured that this growth in the slave trade was the case for the expansion of plantation crops, particularly sugar, tobacco, and coffee, while it was not the case, for example, of the important and profitable cod exports from Newfoundland to Western Europe.

The export from the Americas of plantation crops, and the export of Western manufactured goods to the Americas and Africa that it helped finance, played a major role in restructuring much of the Western economy, and notably that focused on the Atlantic seaboard. These trades powerfully developed and accentuated the role of Europe's Atlantic seaboard, and crucially the importance of port cities, particularly Bordeaux, Bristol, Liverpool, and Nantes. The import of plantation crops also greatly affected the material culture of Westerners, and their diet and health. By supplying new products, or providing existing ones at a more attractive price or in new forms, this trade both satisfied and, in turn, stimulated consumer demand.

Transoceanic trade provided Westerners with goods designed to stimulate the body, mind, and senses: sugar, tobacco, and caffeine drinks. The last were tea (from Asia), and coffee and chocolate (from the West Indies).[1] As none of these goods were "necessary," this was very much consumerism, and, alongside a fundamental change in taste, fueled the start of the modern culture of material fashion with the rapid rise in popularity of passing trends. The fashion for these goods was linked to particular changes in taste as part of a broader-based growth in demand. Sugar came to be much more

DOI: 10.4324/9781003457923-3

important to the response of individuals to food and drink, partly replacing honey as the longstanding sweetener in cooking and drinks. Sugar had been a luxury. When, in 1603, Elizabeth, daughter of James VI of Scotland and I of England, came south from Scotland with her father, she was presented with a sugar loaf at York, which was seen as a sign of respect. This luxury status for sugar ended with increased availability. Growing demand for sugar interacted with rapidly rising supply from the Americas. As a result of the latter, the average retail price of sugar fell considerably in Europe in the second half of the seventeenth century, which, itself, greatly encouraged demand. This demand, in turn, led the trade to become more predictable, which encouraged more investment, and that, in turn, led to further downward pressure on prices. A similar process was seen with personal electronic devices over the last 40 years.

Taste as well as price was at issue. The addition of sugar to drinks greatly increased their popularity by making them easier on the Western palate, while, as the consumption of caffeine drinks surged, so demand for sugar markedly increased. Thus, chocolate was altered by the addition of sugar, making it a sweet rather than a bitter drink. This change made chocolate as a drink more popular in Europe and encouraged the growth in export to that continent of its main ingredient, cacao. Sugar made both tea and coffee more attractive drinks. Sugar was also added to cakes, biscuits, medicine, and other products. The effect was of a major change in the Western palate, and one that helped distinguish the Western world. This change was crucial to the profitability of the Atlantic slave trade.

As with sugar and other plantation crops, the development of cacao production helped drive the slave trade. Much cacao was obtained by the Spaniards from the native population, but this source was supplemented by a plantation production that increased control and predictability. The sale of cacao from the Spanish-ruled Caribbean coast of modern Venezuela began in the 1610s, and encouraged and financed the import of African slaves there: cacao provided the necessary demand and money. This cacao was sold to Spanish markets in the Americas, which is a reminder of the variety of the Atlantic economy in which slavery played such a dynamic role: trade was not simply across the Atlantic. The varied trade routes increased the profitability and durability of the Atlantic economy, and thus the value of investing in ships and slaves. In the 1630s and 1640s, rising cacao sales from Venezuela produced profits and the prospect of more that helped finance and encourage larger slave imports there. Each adult slave earned on average per year about 40 percent of their market value, which was a very high rate of return. The slaves were initially provided by the Portuguese, but, from mid-century, the Dutch, based on the nearby Caribbean island of Curaçao, became more important as suppliers. The mobility of the slave gangs helped expand the frontier of production and, by 1744, there were over 5 million cacao trees in the Caracas province. In contrast to the control wielded over slaves, the *encomienda* system in Spanish America and the employment of the Native American population it offered was of limited use in satisfying new labor demands and, in particular, the scale and control required.[2]

In a process also seen with other products, other countries followed. The French established cacao plantations on their colonies of Martinique and Guadeloupe in the West Indies in the early 1660s, the Portuguese following in Brazil in the late 1670s, and the Dutch (in Surinam) and the English (on Jamaica, but with limited success) in the 1680s. As the production of cacao increased, prices fell, and greater availability at lower prices encouraged consumption in Europe.[3]

The production of sugar from cane was on a greater scale than that of cacao, reflecting its range of uses and the addictive attraction of sweetness. Sugar was profitable, particularly when special opportunities beckoned. These opportunities were usually a matter of war between producers. For example, as on the English colony of Barbados in the 1650s, profits of as much as 40 or 50 percent reflected the impact of the lengthy mid-century war in Brazil between the Dutch and Portuguese on competing Dutch and Portuguese sources of sugar imports into Europe.[4]

Sugar meant slaves. Initially, settlers in the English West Indian colonies had largely been laborers provided by contracts of indenture, a practice of labor provision and control transplanted from England. Plantation work, however, was hard, ensuring that labor availability and control were key issues. The labor regime in sugar and rice cultivation was particularly arduous and deadly; that for tobacco and cacao was less so. Hacking down sugar cane—crucial to the production process—was backbreaking work and even worse under a hot sun, as I found, even though a well-nourished individual with plenty of water available, when practicing hacking sugarcane on Antigua. Although a workforce with origins in Africa would be more used to the heat than Europeans, in the sugar-cane plantations there was not the natural forest cover found in Africa. The bulky and fibrous cane is not easy to work.

Sugar required a large labor force, and slavery provided this more effectively than indentured labor, which was not only less malleable but also less attuned to the environment in the West Indies, in particular the climate. The impact in the late 1640s of disease on the white settlers encouraged this process.[5] Captain William Freeman, who from 1670 developed a sugar plantation on the English island colony of Montserrat in the West Indies, claimed that "land without slaves is a dead stock."[6]

The South Atlantic

All too often the slave trade is seen in terms of the North Atlantic, that from West Africa to the West Indies and North America. However, the South Atlantic trade was crucial and should be discussed first, not least in order to challenge the focus on the North Atlantic. Sugar production developed at a major rate in Portugal's colony of Brazil from the 1570s, producing a "white gold" economy. This economy was based on slave labor. The slaves initially came from Senegambia in West Africa, the part of sub-Saharan Africa where the Portuguese had first arrived. However, the slave trade increasingly linked Brazil to the Portuguese colony of Angola, which is further south in Africa. Sugar production helped ensure that Brazil, which was the largest individual colony, received 42 percent of the slaves imported into the Americas during the seventeenth century, the largest individual flow by colony. The number of slaves who arrived in Brazil exceeded that of white settlers.

This flow was necessary because, on the sugar estates in Brazil, slaves had an average life expectancy of up to eight years only. This was a grim reality that reflected the arduous nature of much of the work there and the harshness of most of the conditions, a grim reality that left scant place for individual narratives. At the same time, this average period was long enough to make the purchase of slaves profitable in terms of the value they produced.

In Africa, Afro-Portuguese (Luso-African) slaving networks, rather than the Portuguese government or Portuguese trading companies, provided the necessary capital to buy the slaves, dominated the supply, and took the profit. These networks linked the African interior to the Atlantic coast, both in Guinea (West Africa) and in Angola. The Portuguese presence expanded, as in 1616 when a new port in Angola was opened at Benguela in order to support the trade and expand the area on which it could readily draw. Benguela was to take second place to Luanda, the original Portuguese base, as the port for the Angolan slave trade.

The Dutch attempt to conquer Brazil from the Portuguese from the 1620s, however, greatly disrupted sugar production there, not least because the Dutch focused on north-east Brazil, the most accessible and profitable part of the colony, as well as the part closest to the United Provinces (Dutch Republic). The Dutch had considerable initial success in their war, capturing the major port of Recife in 1630. The resulting disruption of sugar production in Brazil led to a marked shift in sugar production to the West Indies. This process was accentuated when the Dutch failure in Brazil, which began in 1645, was followed, in 1654, by the Portuguese recapture of Recife, a key success. This recapture was consolidated by the expulsion by the triumphant Portuguese colonial authorities of Dutch and Jewish settlers from Brazil. They, in turn, brought to the Caribbean their capital, mercantile contacts, and expertise in sugar-mill technology. This was a significant transfer as, earlier, that of the Genoese and Venetian finance and entrepreneurial drive from the Black Sea to the Atlantic had been in the fifteenth century.

On the other side of the Atlantic, the Dutch had also captured the Portuguese African slaving bases of Luanda (Angola), Benguela (Angola), and São Tomé in 1641. As a demonstration of the impact of the slave trade and its role in the interaction of Africa and the New World, these captures affected the availability of slaves in Brazil. The resulting shortage encouraged Portuguese slave hunting in the Brazilian interior, which was a less predictable process and without the supply (a ghastly term for a brutal reality) produced by Africans enslaving other Africans. In 1648, however, a Portuguese fleet from Brazil recaptured these bases and restored the trade.[7] Thus, the fate of slaves and those who might be shipped across the Atlantic was very much affected by the vagaries of war. Aside from the direct impact of military operations, war greatly influenced the economics of trade, not least in terms of the risk and costs of disruption.

Portuguese resilience in the South Atlantic (a resilience not matched in important parts of Portugal's Asian empire that was attacked by the Dutch in mid-century, notably Sri Lanka and Malacca) provided the basis for a marked revival in the second half of the seventeenth century of the integrated Portuguese slave and sugar economy in the South Atlantic. The profitability of Brazil, in turn, meant that the Portuguese Atlantic empire did not suffer from the lack of capital and relative uncompetitiveness seen in Portuguese Asia. This situation ensured that the South Atlantic slave trade could serve to accumulate capital and provide an opportunity for fresh investment. This resilience was also important to the development of the British Atlantic, as Portugal was to prove a key economic and financial partner to Britain, and notably in the first half of the eighteenth century.

Dutch slave exports to Brazil were affected by the Portuguese victory. As a result, the Dutch, instead, increasingly focused on Spanish American markets (rather than Brazil), using their Caribbean bases of Curaçao and St. Eustatius as entrepôts, particularly to supply nearby Venezuela. The Dutch also supplied slaves to their own colonies in the Guianas on the northern coast of South America: Essequibo, Demerara, where New Amsterdam was founded in 1627, and Surinam, where Paramaribo was founded in 1613. The Dutch competed with England on this coast, especially in Essequibo and Surinam, while further east, Cayenne, the basis of French Guiana, was founded as a colony in 1635. At that stage, France was an ally of the Dutch against Spain, although alliance and enmity did not always determine slave trading, not least due to the major role of smuggling. In turn, smuggling increased the dimension of risk in slave-trading.

The West Indies

The French, English, and Dutch also all acquired bases further north. In addition to Cayenne, the French settled a number of islands in the West Indies—St. Christopher (St. Kitts, 1625), Martinique (1635), Guadeloupe (1635), Dominica (1635), Grenada (1650), and St. Domingue (now Haiti, 1660)—as well as Louisiana. Claimed by La Salle in 1682, the latter had its first French base at Fort Maurepas in Biloxi Bay on the Gulf of Mexico coast of the modern state of Mississippi, in 1699. The French also established positions on the coast of Africa: St. Louis in 1638, Gorée in 1677, and Assinie in 1687 all becoming bases for slavers.

The British—at this stage the English—established settlements on Caribbean islands that the Spaniards had not colonized, although this was not an easy process. Settlements were founded on St. Lucia in 1605 and Grenada in 1609. However, opposition from native Caribs helped lead to their failure and provided a valuable instance of the folly of assuming an automatic Western military superiority with supposedly clear-cut consequences. That point is more generally true for the Atlantic world. European powers were unmatched at sea, but on land there was a persistent need to adapt to a very different reality.

Such an adaptation, however, was less necessary on small islands affected by the presence of Western naval power and where there were many settlers.

Bermuda, an island in the Atlantic remote from other islands, was discovered in 1609, probably helping inspire Shakespeare's play *The Tempest*, although it has also been claimed that there were alternative sources of inspiration. Settled in 1612, Bermuda became a successful colony where tobacco cultivation was swiftly introduced. The first black slaves were taken there in 1616. Bermuda was followed by the establishment of lasting English colonies on the Caribbean islands of Barbados in 1627, Nevis in 1628, and Antigua and Montserrat in 1632, although an attempt upon the Spaniards on Trinidad in 1626 failed.

These colonies were no mere adjunct to the English possessions in North America. Instead, they generated more wealth and, until the 1660s, attracted more settlers than the possessions on the North American mainland, Barbados proving the most popular destination. Originally settled in 1627 by colonists funded by William Courteen, a London merchant, Barbados focused first on tobacco and then expanded into cotton and indigo. The plantation system then adapted to sugar. Colonies that shifted from tobacco—the price of which slumped in the 1640s—to sugar, in effect another drug, saw a marked increase in the slave population and less of a reliance on indentured labor: on Montserrat, 40 percent of the 4,500-strong population in 1678 was nonwhite, but this grew to 80 percent of the 7,200-strong population in 1729.[8] The same process occurred on other islands such as Barbados, which was hit by a falling supply of white indentured servants. These servants were now attracted, instead, to the English mainland colonies of South Carolina, Virginia, and Maryland, where prospects were better, not least for acquiring land.[9]

This shift drove the slave trade to the English West Indies. Slaves suddenly appeared in Barbados deeds in 1642 as a result of the arrival of English slave ships there the previous year. Although prices of slaves thereafter fluctuated annually, the trend was one of a fall over time.[10] This fall reflected the commodification of humans that was so central to the slave trade. The economics of mass production were applied to human beings and these economics ensured that, as the price fell, so demand rose and supply was increased. Conversely, it might have been increased supply from Africa, rather than colonial demand, that drove lower prices and thus encouraged a shift to slaves as opposed to indentured labor. At any rate, the consequences at the individual level were horrific.[11]

Falling prices both encouraged the trade and reflected its more sophisticated organization, which was at once responsive to both sources and markets. Price, however, was not the sole factor in encouraging the use of slaves. Slaves offered a longer labor availability than that provided by indentured servants.[12] Initially, a colony with only a small minority of Africans, about 200 out of 6,000 people in 1637, Barbados had a majority of black people by 1660.

The English presence in the West Indies was expanded by the settlement of Jamaica and the Cayman Islands from 1655, the Virgin Islands from 1666, and the Bahamas from 1670. However, these were islands that had not hitherto attracted much European interest and several appeared marginal economically. Indeed, much immigration by both white migrants and slaves was, instead, to already established English colonies such as Barbados. In turn, land-hungry Barbadians settled elsewhere, notably on Jamaica. There, they sought to subjugate the African slaves of the previous Spanish settlers, slaves who had used the English conquest as an opportunity to win freedom. The size and forest cover of Jamaica made it impossible to control. The spread of sugar production in Jamaica from the late 1660s was linked to the increase in slavery, both the hunting down of escaped slaves and the import of many new ones. Whereas in 1662 there were about 3,500 white people and 550 black people in Jamaica, by 1673 they were each about 7,700, and by 1681 there were 24,000 black people and 7,500 white people. Over 1,800 slaves were being brought to the island each year.

Buccaneering and the contraband trade remained important in the English colonies, helping indeed on Jamaica to provide much of the initial capital that financed plantation agriculture.[13] This was an aspect of a more general overlap between the slave trade and illegal activity, an overlap that was not so common with slavery itself. This overlap helps ensure that discussion simply in terms of the regulated trade or what "ought" to happen does not take us particularly far.

More settled activity also developed on the slave islands where land was rapidly cleared and used for commercial agriculture, although much remained uncleared, notably mountainous terrain. Indeed, plantations were often close to land that remained uncultivated.

The labor-intensive nature of the plantation economies led to a need for workers. Sugar was to lead to slaves, or, at least, more slaves and, in particular, to the replacement of tobacco smallholdings by sugar plantations; but it is important not to see this as the inevitable economic and social pattern of the colonies in the West Indies. On Jamaica, the largest English colony in the West Indies, and the one with the most diverse economy and varied agriculture, a mixed economic pattern that was less capital-intensive than that on other English colonies, especially Barbados, was initially dominant, and cacao was the first key plantation product. This mixed pattern continued to be important even after there was an emphasis on sugar, although sugar dominated exports from Jamaica.[14]

By 1675, there were 70 plantations in Jamaica able to produce 50 tons of sugar each year. On Jamaica, the black population rose to 42,000 in 1700, when there were only 7,300 white people. Jamaica largely switched from smallholdings to plantation monoculture. During this period, slave buying was widespread among the white community, but large purchasers dominated the market, which reflected their access to credit. In turn, the role of large purchasers accentuated social stratification among the white population. The market in slaves became more complex and controlled by specialized traders, with a growing resale market within Jamaica which further accentuated the cruel instability of the slaves' lives, one in which enforced mobility and the accompanying social disruption played key roles.[15] Sugar was not the sole plantation crop in the West Indies in the late seventeenth century: on Jamaica, the English also had cacao, cotton, ginger, and indigo plantations, each worked by slaves. Similarly, the use of slaves in agriculture elsewhere in the New World was not simply linked to export agriculture, while some slaves were not involved in agriculture.

Sugar was more profitable than tobacco and cotton, as well as being arduous to cultivate. This drove a change to the economy, society, and demographics of the Caribbean islands. Greater profits meant more money to spend on slaves, and slave labor transformed the society and culture of the islands. The labor regime was far harsher than that of indentured servants, harsh as that was. This harsher regime, which reflected the absence of slave rights, the possibility of brutal control, and racism, a regime marked on slave bodies by scars, encouraged a violent response by some slaves, with flight, conspiracy, and rebellion. Although less harsh, the treatment of indentured labor also led to a similar response.

In turn, opposition resulted in legislation that imposed strict controls and sought to prevail by intimidation. Initially, the legislation largely treated servants and slaves in a similar fashion, but slaves were increasingly distinguished, being subject to a separate code from the Barbados Assembly in 1661, in contrast to the more piecemeal situation in 1652. The entire stance was far harsher. The preamble to the Barbados Slave Act described "Negroes" as a "heathenish brutish, and an uncertain dangerous pride of people" who required harsher "punishionary laws." The word "Negro," meaning black African, was used interchangeably with slave, while description as "brutish," a word associated in English with beasts, further categorized and disparaged. "Pride," a word employed to describe a band of lions, further treated slaves as animals with the additional sense of being brutish, fierce and dangerous. As such, slaves could not possess rights, in contrast to the position taken in Spanish and Portuguese laws and to be taken in the French *Code Noir*. A clear legal difference was established in the Barbados Slave Act, as slaves were not to receive a jury trial of twelve peers, but rather to be tried in what was to be known as a slave court, in which two justices of the peace and three free-holders (all five inevitably white) were to pass judgment. Severe punishment was to be inflicted on any "Negro" who struck a "Christian," a definition denoting a white person, but also drawing on religious justification. The first offense was a severe whipping, while, for the second, the offender was to have their nose slit and face branded, a brutal disfigurement to demonstrate status and punishment. The killing of a slave during punishment was not to be considered a crime. In contrast, the legislation of 1661 protected European indentured servants from such violence. A clear racial difference was affirmed, and this difference made harsh treatment easier to justify and impose.

Harsh legislation was linked to the demographic situation. The numerical relationship between numerous slaves and fewer white people led the latter to support garrisoning by soldiery; the removal of soldiery for whatever reason alarmed the white population. White people felt threatened by the prospect of insurrection from the plantation slaves, as well as by possible attacks by runaway slaves—the latter a particular threat on Jamaica. The sense of alarm was communicated to Western readers, in Europe and the Americas, by publications such as *Great Newes from the Barbadoes, or, A True and Faithful Account of the Grand Conspiracy of the Negroes against the English and the Happy Discovery of the Same* (1676). In 1684, the Jamaica Assembly passed a new Slave Act that sought to codify slavery by regulating the control of runaways, the obligation to enforce slave law, and property rights in slaves.

These rights were changed to increase those of creditors: if owners died in debt, slaves must be sold in order to satisfy their creditors. This provision helped liquefy slave property, but moving slaves from landed estates in this fashion added a further level of disruption to the fact that they were treated as property. Slavery was thus to be characterized by high labor mobility, rather than as part of a stable, albeit highly coercive, pattern of settled agriculture.

The expansion of plantation crops and the slave economy took place within an economic system made dynamic by competition and a territorial system driven by conflict and the prospect of conflict. The value of plantation exports and the vulnerability of the Caribbean islands to attack encouraged the European powers to try to seize each other's positions, although they were frequently unsuccessful. Thus, during the Franco-Dutch war of 1672–8, the Dutch failed to take Martinique from the French in 1674, and the French Curaçao from the Dutch in 1678. The English Western Design of 1654–5 against the Spanish colony of Hispaniola failed, and Jamaica appeared a meagre consolation prize. At war with France from 1689 to 1697, English forces also unsuccessfully attacked Guadeloupe in 1691 and Martinique in 1693, and failed to capture Saint-Domingue in 1695. This dynamic character again undercuts any notion of a normal or predictable system, and underlines why what "ought" to happen in terms of the regulated trade only takes us so far.

West Africa

Increased demand for slaves in the New World had meanwhile accentuated and refocused European interest in West Africa. However, trade there was not easy. For example, the Company of Adventurers of London Trading to the Parts of Africa (the Guinea Company), which was granted a monopoly by James I of England (James VI of Scotland) in 1618, only traded to the River Gambia in 1618–21 before abandoning the unprofitable trade. Under Charles I, a Scottish Guinea Company that operated on the Gold Coast (coast of modern Ghana) was founded in 1634, but the company, which, in fact, was largely London-based, had only limited success.[16] In turn, the total overthrow of Crown authority, as a result of the English Civil War (1642–6) and its Scottish counterpart, challenged the monopoly rights that rested on it, and the Guinea Company lost its monopoly of the trade on the Gold Coast. On that coast, factories (slave trade bases) were established at Anomabu (1639) and Takoradi (1645), while interloping merchants who were not part of the company were active; another factory was founded on the Benin Coast.

There were also struggles between states over the control of trade from West Africa, a prize made profitable in particular (though not only) by the slave trade. In 1658, the Danish crown, then at war with Sweden, provided backing for the seizure of the Swedish bases in West Africa, including the fort, which was then sold to the Dutch West Africa Company. In 1661–4, there was a bitter conflict at a greater scale. The (English) Company of Royal Adventurers Trading into Africa, chartered in 1660 as the monarchy was restored in the person of Charles II (r. 1660–85), fought the Dutch West Africa Company. In early 1661, the English company sent out a small expedition using royal vessels under the command of the aggressive Captain Robert Holmes, a protégé of James, Duke of York, brother of Charles II and, later, James II (r. 1685–8). The Lord High Admiral, James, was also a prominent shareholder in the company. This was the sort of overlap between government and the "private sector" that set the tone for much Western activity in the Atlantic world, including the slave trade. Holmes seized two islands in the mouth of the River Gambia and attacked nearby Dutch forts, but the Dutch reacted sharply, taking English ships and in June 1663 capturing the key English base at Cape Coast Castle. In November 1663, Holmes was sent to support the company and to maintain the rights of English subjects by force, and, in early 1664, he seized the major Dutch settlements on the Gold Coast. However, a Dutch fleet under de Ruyter then captured the African settlements. Regional competition helped trigger the Second Anglo-Dutch War in 1665, and was subsumed into it.

Such shifts in control greatly affected the local African elites who were dependent on the slave trade. At the same time, these elites sought to play off the foreign powers which helped given the Africans greater agency. The conflicts of the European powers require discussion because they underline the extent to which the situation was not a case of the West or Europeans versus the Africans, as might be suggested if the emphasis is on Western racism or trade or firearms. Instead, that element coexisted with a more complex interaction between competing Western/European powers and their African counterparts.

The conflicts also led to significant changes in the Western world that were to be important to the geopolitics of the Atlantic, and thus of the slave trade. In particular, the Anglo-Dutch Wars, which were formally from 1652 to 1654, 1665 to 1667, and 1672 to 1674, but, in practice, more lengthy outside Europe, left England not only as an increasingly powerful maritime state, but also with a far stronger position in West Africa. Linked to the wars, but also a more long-term process, England's new bases included Cape Coast Castle in 1652, Tasso Island in 1663, Fort James in 1664, and Accra, Apollonia, Elmina (the one-time Portuguese base of São Jorge da Mina), Winneba, and Whydah in 1672.

The foundation of bases reflected not only the struggles of Western powers, but also changing opportunities on the African coast. In the late seventeenth century, there was a rise in the relative importance of slaves from sources from north of the Equator, as opposed to from Angola. In large part, this rise was as a reflection of the greater *relative* significance of the West Indies as a market for slaves, as opposed to Brazil (which, however, remained the greatest individual market), and of the increased role of English and French capital, although Angolan, Brazilian, and Portuguese capital also remained crucial to the Atlantic slave trade. The Bight of Benin, where Anecho became a Portuguese base in 1645, and Whydah an English one in 1672, was of particular importance for slave exports from West Africa. By the end of the century, this area of significant slave exports had extended to the Gold Coast (modern Ghana). In contrast, the Bight of Biafra, further east, and the Sierra Leone coast, further west, did not become important as major sources of slaves until the mid-eighteenth century.

These geographical shifts underlined the importance of developments in African trading networks and in the politics of rival African states, as well as the interaction of those independent or largely autonomous factors with the opportunism of Western merchants. To give agency solely to one, African or Western, would be inappropriate. There is scant sign of much complaint on the part of African leaders about the slave trade; they were perfectly able to stop it if they wished. The politics of African countries, both internally and among themselves, went beyond simply serving as agents of European merchants, as they are so often represented. Slavery was important internally because of the struggles over revenue, and concomitant political structures often involved the use of slaves for the "outs" fiscally to find their way back "in', or for a self-made entrepreneur to invest in slaves as workers and even, if necessary, as soldiers.

Internationally, wars were not necessarily simply slave raids. Wars moreover could be lost or rendered profitless. Dahomey, a key slave raider, lost many conflicts and in others did not capture many slaves. So, while achieving booty was certainly a by-product of conflict, war would usually have another and more understandable political motivation. These came foremost, not the slave trade as a cause of warfare. In economic matters, there was also African agency with demand for Western goods including those from Western colonies, such as tobacco from Brazil, as well as Indian textiles moved by European traders.[17]

International Competition

The prospects of the slave trade encouraged many merchants and numerous rulers to enter into it. At the beginning of the seventeenth century, Spanish pressure had helped deter the Grand Duchy of Tuscany from persisting with plans to create colonies in Sierra Leone and Brazil, while the Duchy of Courland (in modern Latvia) was unable to sustain its bold transoceanic plans. However, Denmark, Sweden, and Brandenburg-Prussia all established bases in West Africa, although the Swedes had lost all of theirs by the end of the 1650s.

Greater Western demand for plantation goods led to, and financed, an increase in the number of slaves imported into the Americas in the seventeenth century, both compared to the previous century and during the century itself. The trend was markedly upwards. About half a million slaves were imported in the first half of the century, but a million in the second half, including over 600,000 in the last quarter, and that in a century which did not see significant population growth. The profitability of the trade encouraged international competition.

The slave trade had initially been dominated by the need to supply the Portuguese and Spanish colonies with labor, but, as the Dutch, French, and English expanded their colonial presence, so they played a more direct role in the trade, selling to their own colonies and not, as hitherto, taking the role of simply selling to the Portuguese and Spanish colonies. The first slaving voyage from Bordeaux, which, with its protected anchorage, financial strength, and economic resources, was to be a major base for the French slaving trade, departed in 1672, although the French slave trade only essentially became important after 1713. Much of the seventeenth-century French slave trade was by interlopers, and thus clandestine. However, whether clandestine or not, France's wars with the Dutch in 1672–8, and with the English and Dutch in 1689–97 and 1702–13, hit its trade hard. In particular, French ships suffered from English blockades, for after a major English naval victory at Barfleur in 1692, England was the major Atlantic naval power and French trade therefore vulnerable.[18]

The English Role

The English Company of Royal Adventurers Trading into Africa, which was re-formed as the Royal African Company in 1672, was granted by its charter monopoly rights over the English slave trade between Africa and the West Indies. However, the overthrow of the Stuarts in the person of James II (and VII of Scotland) in the Glorious Revolution of 1688–9 that brought William III of Orange to the throne, resulted in a decline in government support. This undermining of the company's position hit the company's finances and led to a decline in its assertiveness. This led, in 1698, to the company, in a key policy change, licensing private traders in return for a 10 percent tax. The company's monopoly had become more unpopular due to its failure to meet rising demand for slaves in the New World, and because of the mood and move against monopolies after the Glorious Revolution. This freeing of the African trade in 1698 legalized the position of interlopers (illegal traders) who now could become private traders, although some preferred to continue as interlopers. The private traders took over much of the activity of the company. The role thereby granted to private enterprise helped propel England to the fore in slave shipments as it provided plentiful opportunities for entrepreneurs.[19]

The use of the term "freeing" and the very idea of a "free trade" appears bitterly inappropriate, given that this trade was in slaves and serves as a reminder both that the vocabulary and ideology of economics do not match those of modern human rights, and that understandings of economics and human rights are culturally dependent and change through time. This is not an automatic process, nor one free from debate and controversy, and these affect scholarship which cannot exist in a vacuum, nor without being affected by changes in the use of language and differences in them.

Demographic factors continued to place a heavy emphasis on slave labor. Much higher death rates in the West Indies than in North America, and higher death rates for white people than for slaves, ensured that the colonies there did not become settler societies with large, locally born white populations. Yellow fever, which first struck in 1694, was to be a particular scourge and was especially virulent in white people previously unexposed to the disease, while malaria was a serious problem. In part, disease was due to the slave trade itself, as the ships that brought the slaves carried mosquitoes that had the yellow fever virus. In addition, the cutting down of forests to clear land for sugar cultivation and to provide fuel for sugar refining reduced the birds that were predators of mosquitoes. This cutting down of the forests continued into the nineteenth century, being seen then in Cuba as sugar cultivation spread eastward in the island. This cutting down of trees produced major environmental changes in the Caribbean, not least a marked lessening in biodiversity. Moreover, clay pots used for sugar refining, once discarded and filled with rainwater, became breeding grounds for mosquitoes, and the latter also fed off sugar. Thus, the slave economy was destructive to many white settlers, a pattern that matched the moral degradation involved. Of course, the plight of the slaves was far worse.

Indeed, the greater ability of black people to adapt to the American tropics, not least their stronger resistance to yellow fever,[20] was seen by Westerners as justifying their use for hard labor, not least because it was argued that this adaptation reflected the extent to which they were like animals.[21] Slave labor was one of the ways in which Westerners sought to respond to, and profit from, the tropics. The harshness and brutality involved were not solely a reflection of racism, important as that was, but also of an instrumentalism within the technology and opportunities of the age. As an instructive guide to change (but one without the human cost), a modern equivalent would be the use of machinery, particularly of air conditioning.

In North America, the British established their first permanent colony at Jamestown on Chesapeake Bay in 1607. The colony expanded as a result of the continued arrival of new settlers. A key opportunity for profit arose when tobacco became the major crop in both Virginia and Maryland, or rather both colonies near the bay. The need for labor there and the long terms of service exacted in return for transportation to the New World encouraged a dealing in servants, and the hiring of servants was harsher and more degrading than in England.[22] This use of white labor was accompanied, from 1619, when the first shipment arrived, by that of African slaves.

Tobacco's limited capital requirements and high profitability fostered investment and settlement, while, because tobacco was an export crop, the links with England were underlined.

The needs and difficulties of tobacco cultivation and trade, however, created serious problems for farmers, and this situation ensured particular sensitivity to labor availability and price, as that was the easiest cost factor to affect. Prior to the 1680s, savings in the costs of production and marketing were important in expanding the market for tobacco. Then, there were about three decades of stagnation at a time of rising labor prices, followed, during the eighteenth century, by increased demand, although also rising production costs. Moreover, there were short-term booms and busts, as well as a shifting regional pattern in tobacco production. These changes affected the distribution of slaves and the direction of the slave trade.

There was a strong move to slaves in the tobacco-based Chesapeake labor system in the late seventeenth century. White indentured workers were difficult to retain in the face of the opportunities offered by rapidly spreading English settlement. As so often in the history of the slave trade, demographics proved a key element; in this case the fall in the birth rate in England in the 1640s. This hit Chesapeake planters in the early 1660s, with both fewer young men entering the Virginian labor market from England and a rise in real wages in Virginia. These difficulties encouraged the move toward slaves, as labor costs thereby could more readily be fixed.[23] Moreover, slaves could be legally distinguished from indentured servants, as with the legislation passed by the Barbados Assembly of 1661.

Captured Native Americans were used as slaves—for example, they were sent from Connecticut to Barbados after King Philip's war in 1675–6, while Tuscaroras defeated in Carolina in 1715 were enslaved, although they were insufficient in number. Africans were also regarded as more effective and industrious workers than Native Americans, and less likely to escape successfully, and, accordingly, commanded higher prices. This was true not only in Brazil, but also in eighteenth-century Carolina (where there were most Native American slaves), and more generally. The profitability of the plantation economy made it possible to invest in these African workers, and it was desirable to do so.

The development of slavery interacted with white racism in a context in which the English had to develop legal codes and practices to match a labor institution that was new to them. Slavery, racism, and law were all linked in the development of a colonial society that used the slave labor already seen in the Spanish and Portuguese colonies, but with distinctive English legal and political structures. Economic advantage and coercive power were related to a belief that Africans were inherently inferior, an attitude that drew on what were seen as their innate characteristics and the fact that they had arrived in the New World as slaves.[24] The latter meant that they were regarded in a harsh light, which was a classic instance of hardship being visited on the victims.

The establishment of Carolina as a separate colony in 1663 expanded the English presence in North America southward and led to a marked rise in the slave economy of the English Atlantic. The new colony was closely linked to the English West Indies. Key figures included Sir John Colleton, formerly a Barbados sugar planter, and Anthony, 1st Earl of Shaftesbury, who had invested in Barbados. The Corporation of Barbados Adventurors stressed their position and value as "experienced planters." The Fundamental Constitutions of Carolina of 1669, which were probably drawn up by Shaftesbury's secretary, the Whig philosopher, John Locke, promised that every freeman should have "absolute authority over his Negro slaves," and made it clear that slave conversion to Christianity would not alter this relationship. There was a repeated concern that conversion would automatically free slaves. Carolina provided opportunities for younger sons from the crowded islands, especially Barbados. The settlers from these islands brought black slaves with them, ensuring that a sizable black labor force was soon established in the new colony. In 1691, the South Carolina Assembly modeled its slave code on the harsh Jamaica Slave Act of 1684 that was passed in response to slave resistance on the island. Carolina became a source of supply for food to the West Indies, as well, eventually, as a key exporter of colonial goods, including, from the late 1690s, rice. Rice cultivation required large numbers of slaves, as did that of indigo.[25]

Nevertheless, although developments in the late seventeenth century were significant, much of the growth in the North American slave economy was to follow in the eighteenth century as that economy developed and as an important part of a growing Atlantic economy. Within the latter, the crossing distance, and therefore time, from Africa to North America, a key index of profitability in the slave trade, was greater than that from Africa to Brazil.

The price for male slaves was higher than that for females, both on the African coast and in the New World, and more males than females were imported into the latter, reflecting the role of demand factors in the trade, specifically the hard physical nature of the work that was expected. In contrast, the tendency in the Constantinople (Istanbul) slave market was for women to cost more, which probably reflected their use for household tasks and sexual roles.

Profit and the prospect of profit in the New World plantation economies owed much to the rights of owners who not only possessed slaves but also controlled the future: the future of their slaves, whose status was inherited as well as personal, and of their estates, the inheritance of which included enslaved people. This continuity meant that owners had a key incentive in getting their slaves to have children. This was literally wealth generation. As a result, concerns about interracial unions were a matter not only of racial and gender control, but also of economic value. These unions demonstrated the exploitative economy and unequal relationships of slavery: owners enslaving the children of their partnerships with nonwhite women were a key aspect of the valuation of people for production. This valuation cut across the bonds of family and that also reflected worries about the potential consequences of deracination (the loss of racial identity through biracial children). The dominance of white men over the family life of black women and men was an aspect of the broader moral corruption of slavery, one that drew in part on Classical models of slave ownership. Slaves were not permitted to form families unless owners agreed. The issue reflected not only the views of individual owners but also the controlling strictures of authority, both lay and clerical. Thus, in 1724, the *Code Noir* banned French men and women from living in sexual partnerships with slaves. The use of the term "moral corruption" may appear to pose the danger of using modern standards to judge the situation. However, aside from the point that it is appropriate to understand the past not only in its terms but also for what it can teach the present, it was appreciated by some contemporary moralists that this dominance by white masters could cause not only serious damage to those who suffered, but was also a moral corruption of these masters, in particular by allowing them to indulge sexual control.

The Situation in Africa

It would be wrong to present the complex dynamics of enslavement simply in terms of rising demand for labor in the New World. It is also necessary to look at the African dimension, an issue that generally attracts insufficient relative attention in Western discussion, and notably popular consideration. Accounts focusing on Western economic domination in Africa, on the way in which the Atlantic slave trade encouraged slavery within Africa, and on the gun-slave cycle, by which slaves were obtained by the Westerners in return for the provision of Western guns to the Africans, are inadequate.[26] This is not least because it was not until after the Industrial Revolution had transformed Western Europe's economy in the nineteenth century that traders could exert strong economic pressure on Africa, while Western weapon sales, although important in the eighteenth century, did not provide the key to the trade. Instead, it is necessary to focus on the supply of African labor and nature of African agency,[27] and thus on the means for satisfying Western demand, and also to offer a specific examination of the different slave-supplying regions. Such an examination is made more complex by the widening impact of Westerners along the African Atlantic coast in the seventeenth century, although the Bight of Biafra to the east of the Niger delta was not to become an important slave-supplying zone until the eighteenth century.

What emerges clearly in discussing supply is a politics of frequent conflict within Africa that produced slaves, and of the rise and fall of states, such as the collapse of the Mali Empire in about 1660. Fighting in Africa was also often linked to serious droughts and famine, although the introduction of maize, manioc (cassava), and peanuts from the Americas helped support population numbers. Droughts and famine increased conflict and indebtedness within Africa, and both led to more slavery. The seizure of people for slavery was seen as a way to weaken rivals—certainly so in Senegambia and the Gold Coast. This was true both at the international level, where war captives were enslaved, and within states, notably when people were pawned in order to repay debts.

The availability of large numbers of slaves helped depress their price in West Africa, which meant that their purchase was more easy and efficient as a way of addressing economic needs in the New World, a sentence that conceals much misery. This economic context was furthered because the nature of African agriculture, which emphasized the use of the hoe, was relatively inefficient. Inefficient agricultural practices lowered the value of African labor, thus increasing the importance of selling slaves, while also depressing the cost of slaves and, in turn, thereby encouraging their purchase.[28]

Campaigning was a key aspect of the way in which African states pursued their interests and redefined relations, and slaves resulted from this campaigning. Some states became centralized, while others were decentralized and governed by coalitions of autonomous interests. Each process could owe much to the slave trade, in part depending on how the resulting profit was controlled and distributed. Thus, on the Gold Coast, the slave trade encouraged the disintegration of the small, centralized Fante kingdoms on the coast into a decentralized federation.[29]

Warfare between African powers certainly provided large numbers of slaves. Western powers, in contrast, were not able to seize large numbers. By far the most expansionist Western power in Africa (from the fifteenth century until the early nineteenth when it was succeeded in that role by Britain) was Portugal, but its experience there revealed the major limitations of Western land warfare in Africa prior to the nineteenth century. In South-East Africa, Portuguese attempts to operate from their bases in Mozambique up the River Zambezi and to exploit the civil wars in Mutapa (modern Zimbabwe) were thwarted in the 1690s, with Changamira, the head of the Rozwi Empire, driving the Portuguese from the plateau in 1693. In Angola, the colony from which Portugal obtained most slaves— slaves who were shipped to Brazil—the Portuguese were effective only in combination with African soldiers. Unlike the nineteenth-century pattern of European-organized units filled with African recruits, the Portuguese in seventeenth-century Angola were all organized together into a single unit with its own command structure, while the Africans—either mercenaries, subject troops, or allies—were separately organized in their own units with their own command structure. It was only at the level of the army as a whole that Portuguese officers had command, providing control for entire operations. The Portuguese found the Africans well armed with well-worked iron weapons, as good in some ways as Portuguese steel weaponry, and certainly better than the native wood and obsidian weapons of the New World. The slow rate of fire of Portuguese muskets and the open order (as opposed to densely packed nature) of African fighting formations in Angola reduced the effectiveness of Portuguese firearms, while the Portuguese inability to deploy anything larger than a small force of cavalry ensured that they could not counter this open order. Moreover, their cannon had little impact on African earthwork fortifications which absorbed the shot. In addition, as in North America, firearms diffused rapidly, largely through trade with the outsiders, particularly for slaves, and Africans in Angola possibly even had firearms in equal numbers to the Portuguese already in the 1620s.[30]

Portuguese weakness in Africa, and the strength, in contrast, of rival, non-Western slave traders, was also closely shown in East Africa. Fort Jesus, the well-fortified Portuguese garrison in Mombasa (modern Kenya), fell in 1631 to a surprise storming by Sultan Muhammad Yusuf of Mombasa, and a Portuguese expedition from their major Indian base of Goa failed to regain it in 1632. The Portuguese were able to return when the Sultan abandoned the fortress under the pressure of their attack, but in 1698 Fort Jesus fell again, this time, after a lengthy siege, to the Omani Arabs, who had a significant maritime presence in the Arabian Sea including to the Swahili coast of East Africa. The Portuguese presence on the Indian Ocean coast of Africa north of Mozambique was lost in 1698 until briefly revived in 1728–9. Thereafter, a European territorial presence on the Swahili Coast there did not resume until the 1880s when Germany and Britain established bases.

The development of new military forms and the spread of firearms in West Africa affected Western options ensuring a reliance on purchase not seizure, and they need to be discussed in any treatment of the slave trade, not only because they did so, but also as they influenced the pattern and nature of slave availability. African warfare was transformed by the increasing preponderance of firepower over hand-to-hand combat, and by a growing use of larger armies. From the mid-seventeenth century on, the role of archers increased and missile tactics came to prevail in West Africa, as in Europe, despite the absence yet in West Africa of the widespread use of firearms. The new military methods spread—for example, to Asante in the 1670s.

In turn, the bow was supplanted by the musket. Firearms came into use in Africa over a long timespan, in part because usage was restricted and affected by the limited availability of shot and powder, which reflected the difficulties of providing both as they were dependent on raw materials that were not widely available. The use of firearms in West Africa was first reported in Kano in the fifteenth century, in what is now northern Nigeria. Kano was one of the Islamic Hausa city-states that traded across the Sahara, but a regular force of musketeers was not organized there until the 1770s. On the West African coast, the Asebu army of the 1620s was the first to include a corps of musketeers, their guns being supplied by the Dutch in return for slaves, and muskets replaced bows in the 1650s to 1670s. However, the tactical shift towards open-order fighting in West Africa did not come until later: musketeers were used as a shield for the javelinmen, and tactics centered on missile warfare were slow to develop in the coastal armies. Firepower, nevertheless, increased in the 1680s and 1690s as the more reliable flintlock muskets replaced matchlocks, a change also seen in Europe in this period, although bayonets were not used on the West African coast. The flintlocks were obtained from Western traders. In the forest interior of West Africa, in addition, muskets replaced bows in the 1690s and 1700s.

The emphasis on missile weapons, bows, and, later, muskets interacted with socioeconomic changes, in particular with the transformation of peasants into militarily effective soldiers. This development led in West Africa to the formation of mass armies and to wars which lasted longer. Larger armies increased the numbers who could be captured in conflict, while wars took place over much wider areas, also producing large numbers of slaves. Disposing of captives as slaves weakened opponents. Warfare based on shock tactics had been selective in its manpower requirements, but on the Gold Coast (coastal modern Ghana) and the Slave Coast (coastal modern Togo and Benin, formerly Dahomey), the replacement of shock by missile tactics was linked to a shift from elite forces relying on individual prowess to larger units, although in Dahomey there was an emphasis on a small standing army. These military changes were related to the rise of the states of Akwamu and Asante on the Gold Coast and of Dahomey on the Slave Coast, and maybe to the late-seventeenth-century expansion of Oyo further east. These powerful states were very much to affect Western options, at a time of the major expansion of the slave trade. There was far less opportunity for Western powers to overawe, manipulate, or intimidate local states than there was to be, for example, in southern India from the 1740s.

Unlike Portugal, other Western powers did not try to make conquests in sub-Saharan Africa until the nineteenth century. Instead, the Western presence in West Africa was anchored by coastal forts that served as protected bases for trade, although in some areas there were no settlements and the traders operated from their ships which were moored offshore. These bases were vulnerable to attack: the Dutch position at Offra and the French one at Glehue were destroyed in 1692, the Danish base at Christiansborg fell in 1693, and the secondary British base at Sekondi fell in 1694. In contrast, the leading British base at Cape Coast Castle, the overseas headquarters of the Royal African Company, which was gained in 1652, was never taken and was successfully defended against African attack in 1688. Nevertheless, the British were well aware of the weakness of their position. The garrisons of Western forts were very small and, both for their own security, and for the capacity for any intervention in local conflicts, relied on the forces of African allies.[31]

The situation was only different in South Africa where, benefiting from a favorable environment and from a lack of many native people, Dutch settlers from their base at Cape Town, founded in 1652, established settlements in the surrounding countryside. The Dutch used slave labor which was largely brought from Portuguese Mozambique and from Madagascar, where there was a precariously-established French base.

The Nature of "Cooperation"

As for Westerners in India and the Far East, the emphasis in West Africa was on cooperating with local rulers. British bases in West Africa were not held by sovereign right, but by agreement with local rulers. Rent or tribute was paid for several bases. Moreover, the officials of the Royal African Company sought to maintain a beneficial relationship with numerous local caboceers (leaders) and penyins (elders) through an elaborate and costly system of presents and jobs. Similarly, on the Gold Coast, the Swedish African Company was able to play a role in the 1650s as the Futu elite was seeking a balance to the influence of the Dutch West India Company, just as in the 1750s there was an attempt there to play off the English and the French. The cooperation of the local elite was crucial to the establishment of new trading posts. Cooperation was a matter not only of trade, but also of military and political support.[32]

In so far as comparisons can be made, Western slave traders did not enjoy coercive advantages in Africa greater than those of Arab counterparts on the Indian Ocean coast of Africa. Furthermore, the coercive advantages of Moroccan and other slave raiders operating across the Sahara, and from the *sahel* belt into the forested regions further south, were probably greater, indeed far greater, because they could deploy power by land and use horses.

The major Western advantage, instead, rested on purchasing power in the shape primarily of goods to sell but also benefiting from the availability of bullion-based liquidity stemming from the conquest of silver and gold producing areas in the Americas. This advantage in part derived from the prosperity of plantation economies in the Americas, a prosperity that rested on the integrated nature of the Atlantic economy. Economic factors were crucial. Money, profit, and credit flowed round the system and kept it dynamic, directing the trade to focus on slaves, and not on other products such as gold. However, within West Africa itself, Western Europeans did not have a monopoly of purchasing power as far as slaves were concerned, while in Africa as a whole this was even more the case.

If the emphasis in the discussion of the slave trade is on Western European purchasing power, rather than coercion, then a key element in the slave trade becomes not only the conflicts within Africa that produced slaves, but also the patterns of credit and debt that operationalized and transmitted this purchasing power. African society was thereby opened to demands for slave labor. In this approach, the undercapitalized nature of the African economy emerges as important in creating a reliance on Western credit. In addition, the availability of purchasing power encouraged the sale of slaves, and thus the quest for them. The same, moreover, was the case for other external sources of credit in the shape of Arab slave traders. Existing mechanisms for the ownership and sale of slaves were changed, indeed exacerbated, by foreign purchasers.

Slavery within Africa was already a matter of ethnic, social, and legal difference, with slaves low-ranking individuals taken to serve in another household. While slavery was an inheritable condition, many slaves were at least intended to be short-term slaves: pawned in order to meet a debt that was designed to be repaid so that they should then be freed. This relationship reflected the lack of liquidity in the economy and the importance of work as a means to make payment. Many debtors, indeed, were of high social status.

Pawning, however, took on a different meaning if those pawned were sold on for slavery beyond the Atlantic as this made it difficult, although not impossible, to retrieve them. Moreover, Western slave traders, for whom slave labor was valuable not in Africa, but only if the slaves could be transported across the Atlantic, added a highly unwelcome dynamism by setting time limits on the repayment of debt in order to acquire slaves while ships were on the coast. Human pawns were often fitter than those enslaved within Africa and trafficked for longer to the coast.

Alongside the drive from external demand, local supply and local cooperation in the slave trade were crucial for it to work.[33] African brokers and facilitators were important to a Euro-African society that was dynamic as well as transient, and that dealt in other exports from Africa including beeswax and dyewood. A similar pattern was seen in European slave trading in the Indian Ocean.[34]

Atlantic slavery entailed the interaction of the Western economic order and the dynamics of African warfare, at the cost of the victims of the latter.[35] The demands of the Western-dominated Atlantic economy pressed on local African power systems and, in providing slaves, these systems served the Atlantic economy, even while many slaves were kept for use in the local economy. In a pattern also seen, albeit to a lesser extent, in the Americas, African rulers proved more than willing to sell captives, deriving considerable profit from the trade. In pricing and selling people, Africans responded not only to Western purchasers but also to demands from within African society. This was particularly the case in terms of the provision and pricing of women as opposed to men. Gender politics, social practice, or sexual exploitation—all were variously involved. In West Africa, high rates of polygamy (one man, many wives) in some regions, notably the hinterland, ensured that demand for women and prices for female slaves within Africa were higher in these areas, which affected the number exported, as opposed to the gender difference arising essentially from greater New World demand for male labor. However, as an instance of the difficulty of establishing the reasons for slave flows, it has been argued that this pattern was only a marginal factor.[36]

Trauma and Pain

In the slave trade, the compromises of negotiated shared interests, and of the relations summarized as globalization, were made at the harsh expense of others. It is important to see both elements. Linked to this, it is naïve and limited to treat the slave trade in terms only of compromise, or simply of coercion. Indeed, there was opposition within Africa to both relationships. Two popular religious movements in West Africa, led by Nasr al-Din in 1673–7 and by Abd al-Kadir from 1776, included hostility to the sale of slaves to Christians among their Islamic reform policies. Abd al-Kadir led Muslims in a *jihad* along the River Senegal.[37] These movements failed, but are a reminder of the contentious nature of the trade in contemporary Africa.

The widespread belief among many Africans exported as slaves that they had been sold to cannibals to be cooked and eaten, possibly expressed a wider opposition to the cannibalistic social politics of selling slaves to foreigners. It was also (again mistakenly) believed that Africans were purchased in order to be used as bait for fishing or for making leather shoes out of their skin——brutal displays of power and rumors that reflected the broader mental world that the slave trade gave rise to. Most of those who were taken to the slave ships never returned and it was easy to assume that they had been killed. In African attitudes, there was a linkage of the trade to disease and to death, and reasonably so.

If the "demand" side of the slave trade is morally reprehensible, and not only in modern terms, so also was the "supply" side. The experience of vicious anti-societal warfare in Africa over the last quarter-century, especially, but not only, in Rwanda, Congo, Sierra Leone, Liberia, Sudan, Uganda, the Central African Republic, Nigeria, Niger and Mali including high rates of mutilation, rape, and the slaughter of women and children, has underlined the extent to which foreign intervention in African wars, notably by supplying weapons, is not the sole cause of brutality.

At the individual level, the reality of slavery was of the trauma of capture by African chieftains, of sale to Western merchants, often after passing through African slave trade networks, and of transportation. Each of these involved high levels of violence, shock, hardship, and disruption. Individuals were taken from their families and communities. Many died in the process of capture, although this element of what happened in the African interior is very obscure, as it was also for Abolitionists in the late eighteenth and early nineteenth centuries. Others died in the drive to the coast, in which they were force-marched and joined in coffles or lines, commonly secured to each other by the neck in wooden yokes while leaving their legs free for walking. This method of restraint and control was extremely painful. Again, death rates in this stage are very obscure. Similarly, slaves died in the arduous march across the Sahara to North Africa. Those who fell by the wayside were left to die.

In West Africa, yet more slaves died in the port towns where they were crowded together in degrading and hazardous circumstances while awaiting shipment. At the British base at Cape Coast Castle, the slaves were confined in the vaulted brick slave hole. This was a twilight, stifling, and very unhygienic existence where the slaves were chained round the clock, apart from when they were driven to the Atlantic coast twice daily in order to be washed. Others died on the ships that transported them across the Atlantic, the stage for which death rates have been calculated. They also died soon after arrival in the Americas, as the entire process exposed slaves to unaccustomed levels and types of disease, and in highly infectious conditions. In the process of capture, transportation, and sale, the enslaved were also intimidated, humiliated, and exposed, without any care, to terrifyingly unfamiliar circumstances. The vulnerable were particularly prone to be seized and sold for slavery. Many slaves were orphaned children.

There are no precise figures for overall deaths in the slave trade, but many slaves died on the Atlantic crossing where they were crowded together and held in terrible, especially unsanitary, conditions, with holds proving both fetid and crowded. Furthermore, the slaves had already been weakened by their generally long journey to the Atlantic coast, while there was also an unwillingness on the part of their captors to spend much on provisions. This reluctance exacerbated the health problems already caused by the impact of the conflict that had frequently led to slavery, and of prolonged malnutrition and disease among those who became enslaved. These problems increased the vulnerability of the enslaved, not least to infections. Most died from gastro-intestinal illnesses, such as dysentery. The high death rates were a reflection of the very crowded nature of the ships, the dirty conditions in the holds, and the virulence of infectious diseases. Aside from suicides, the percentage of deaths on a crossing clearly varied, not least because, if delayed, whether for commercial reasons or due to sailing conditions, especially adverse winds, casualty rates rose; but the average percentage was grim: a loss of 17.9 percent on Dutch ships between 1637 and 1645, and the losses for the British Royal African Company between 1680 and 1688, were about 23.5 percent.[38] Statistics record these experiences of brutal custody, for the many who died are now nameless and otherwise forgotten.

This was but a part of a wider process of loss and suffering to which the organization of the slave trade contributed, and which was far greater in intensity than those experienced due to habitual rates of violence in society. The terrible nature of the Middle Passage across the Atlantic, the central element of the slave trade, should not distract attention from the cruelties of the opening phases of the trade in Africa. More generally, as Joseph Miller has pointed out, sequential ownership of the slaves as they were moved meant that none of the owners were responsible for their long-term welfare. Instead, the concern of the owners was to maintain the slaves in a physical condition minimally sufficient to allow quick resale.[39] This need for a minimum condition was important because health and condition were key elements in determining saleability. They were more important than age or gender. Ill, injured, and disabled workers were not wanted. Instead, there was an emphasis on those who could work hard and long, and thus repay the investment represented by purchase and transport. The focus on repayment and on selling fit slaves encouraged speedy sailings once slaves were purchased in Africa. The young were regarded as of particular value as they were readily trainable.

Once arrived in the Americas, the slaves were exposed to fresh difficulties, and renewed humiliation and intimidation. Many slaves were moved considerable distances once in the Americas. Large numbers were transported from island entrepôts, such as Curaçao and Jamaica, to eventual mainland destinations. This process involved fresh voyages and also frequently long marches overland. The former exposed the slaves to renewed crowding and the hazards these entailed, notably from fresh sources of infection. About 15 per cent of the 2.7 million enslaved Africans taken to the British colonies suffered another voyage after the middle passage across the Atlantic, often crowded into holds alongside cargoes, both establishing a distribution system that linked British colonies, as well as the latter to those of other powers. In turn, the return voyages provided goods that strengthened the profitability of the trans-Atlantic trade system.[40]

The marches led to major problems for the enslaved as new ecosystems were confronted, and at a time when resistance to infection was lessened by harsh treatment. Furthermore, it was frequently necessary for the slaves to wade rivers or to climb considerable altitudes.

Becoming habituated to new living and working environments in the so-called seasoning period (an aspect of the commodification of life was the language employed) was difficult and led to a continuation of high death rates. These were accentuated by the conditions of work and life. Poorly clothed and fed, slaves were housed in living quarters that were generally badly ventilated, unhygienic, and crowded. Infestation by fleas and ticks was a major problem.

Poor treatment was set within the context of harsh legal codes. Some were very harsh. Thus, the 1696 Carolina Slave Act decreed that, for the second offense, a male runaway was to be castrated and a woman to lose her ear, and that, if an owner refused to inflict these punishments, he could lose the slave to an informer; while the death of slaves as a result of such treatment was to be compensated, thus encouraging it. These provisions were maintained in the 1712 Act. Accordingly, escapees were castrated. Castration was an aspect of treating slaves like animals, and many slaves would have witnessed such treatment to young bulls and known the pain experienced. This was a cruel management of slaves. Rebel slaves were also castrated in the aftermath of rebellions in Barbados and Jamaica, but there were no similar clauses in their slave laws. Carolina's provisions reflected a strong sense of insecurity in Carolina, a sense that was to characterize the subsequent history of South Carolina. Similarly, in Jamaica, under the 1696 Act, the punishment for a slave hitting an owner was increased to summary execution by the decision of a slave court; those who planned to kill white people could be executed, and the killing or capture of rebellious slaves was to be rewarded. This legislation reflected the major insurrections there in 1685 and 1690, as well as the large numbers of runaways.

The treatment of slaves can be related to the powerful Western racism of the period, with its notion of a clear racial hierarchy, but with cross-currents such as the variety of Christian views on the compatibility of enslavement with Christian doctrine and practice. For many Protestants, Christianity was seen as an aspect of white freedom but others, both Protestants and Catholics, did not share this identification.[41] Western racism was linked to racial classification, to the desire to explain all and to fit all information into a clear system of organized knowledge. In part, this process can be seen as an aspect of the Western Scientific Revolution of the seventeenth century, and notably to the drive to respond to new experimental information, including of the non-Western world. At the same time, the interaction of racism and poor science was significant, as was the classification of evidence in terms of heavily biased social and cultural assumptions. Racism was reflected in Western notions of an inherent hierarchy based on ideas of sharply distinguished races and on supposed differences between them that could be classified in a hierarchical fashion, the genesis of which could be traced back to the biblical sons of Adam. Scientific advancement was not incompatible with false explanations. For example, Marcello Malpighi (1628–94), Professor of Medicine in Bologna and the founder of microscopic anatomy, believed that all men were originally white, but that the sinners had become black. Such self-serving ideas were employed to justify slavery.

Toward a Black America

Brutalized, transported, and treated as disposable goods,[42] the Africans, however, did not simply experience a social death. The deculturation and depersonalization inflicted on them did not define their experience. Instead, for many a sense of self was maintained, a point indicated not so much by the narratives of slaves or former slaves, which are unfortunately relatively few,[43] but rather by other records, including legal ones.

Amid the terrible misery and suffering on the passage across the Atlantic, kinship relations developed between shipmate slaves.[44] Albeit in more difficult circumstances than for slaves within Africa, slaves transported to the Americas were able to rebuild their lives by using, but also crucially adapting, their own experiences and beliefs.[45] In the Americas, Africans developed social and cultural practices that variously reflected African and hybrid forms.[46]

In large part, this process was a matter of survival in a harsh environment. Given the position of the slaves, it was not surprising that participation in the informal economy, including by means of theft, provided a measure of autonomy. Such participation also referred back to huckstering (petty trade) and peddling activities familiar from West Africa. This activity was an aspect of the multi-layered relationship between the slave trade and popular consumerism.[47] African agency and autonomy in the slave colonies were seriously limited. In part, this was because, although there were freed slaves and their descendants, there was no migration of free Africans to the Americas.

However, the dependence of the slave economy on Africans extended to the skills of the latter, as well as the crucial role of some Africans as overseers, the latter an aspect of the multiple divisions within slave society. Slave skills could be fundamental. Thus, the large-scale cultivation of rice in the coastal regions of the British colonies of South Carolina and Georgia in the eighteenth century stemmed from the movement of rice plants and of slaves used to cultivating them from the British slaving bases in West Africa.[48] African cattle herding techniques proved crucial, notably on Barbuda, while canoe construction also drew on West African skills, with the resulting dugouts beneficial both to plantation owners and to the autonomous economic activities of the enslaved.[49] It is also probable that knowledge from the enslaved provided geographical information, for example over the route of the River Niger.[50]

The condition of slaves and former slaves was not simply that of oppression and labor. Differing interpretations of slave life focus respectively on force and adaption (or creolization), although there has frequently been a continuum in interpretations as much as a contrast. Creolization covers both relations between slaves and masters and those among the former, and the range of activities and factors at issue include religion, language, cultural interaction, and economics. For example, the cultivation by slaves of food crops in provision grounds—generally small gardens or strips of land—frequently produced a surplus that slaves used to barter for other benefits within a system of exchange in which, alongside their enslavement, they had agency. In Cuba, the *conucos*, strips of land available, provided goods that slaves sold in taverns, notably pork. More generally, slave agency and exchange could extend to their labor with slaves bargaining over their conditions, albeit within a grossly unequal system. This approach ensures that slave resistance can be seen not simply in terms of the violence and flight that were the counterparts of the force used by slave owners—aspects that are now the focus of attention—but instead, as aspects of a more varied and nuanced situation. The range of settings for slave life and work accentuated this situation. Urban slaves in Spanish America had, for example, reasonable opportunities to improve their position, and freedom was granted there with some readiness.[51] It was granted far more so than in British or French America.

Africans made connections with fellow slaves, a process eased by common experiences in Africa, notably shared language affinities and similar patterns of spirituality. Friendship and family links proved part of this process and, in turn, fostered it. Thus, Africans tended to marry others from the same culture within Africa.[52] This practice helped in the maintenance of traditional beliefs, practices, and rituals, not least by maintaining the relevant knowledge necessary—for example, for devotion to ancestors. Slaves and former slaves could have an active associational and cultural life, with black and mulatto (mixed-race) brotherhoods, for example in Brazil, providing a range of social benefits for members.[53] So also with black churches, as in America prior to emancipation. Moreover, however imperfectly, the courts could provide opportunities to advance, define and protect rights and possibilities, however limited.[54]

Although coercive, the slave trade became part of a dynamic as well as cruel diaspora[55] in which the white people did not, or were unable to, prevent the development of independent associational patterns, patterns that were to prove a key stage in the creation of the black America that was part of the identity of the New World, however much it was underrated or ignored by white people.

Associational networks and values that maintained African or African-style practices and beliefs, for example the Yoruba goddess Yemoja who was a protector of slaves, were matched by others that related more directly to the new world that the Africans, both slaves and free, found. In particular, lay brotherhoods linked to the Catholic Church developed, so that ancestor worship became an aspect of Catholic devotion. This syncretism (merger) matched that by which Native Americans had adapted to Catholicism in the sixteenth century. Moreover, there was important interethnic interaction in the Americas, notably the development of creole languages, such as Papiamentu on the Dutch-ruled island of Curaçao. That language served as a marker of local island identity among the varied diasporic groups that made up the island's population.[56] More broadly, both intermarriage and the nuances of household structures gave slaves an ability to win status and a measure of power. This was particularly so in Latin America, where intermarriage and racial mixing were more common than in the British world.[57]

At the same time, it would be inappropriate to conclude this chapter with that point. For, at every turn, it is necessary to underline that the background and foreground of black existence in the Americas was grim. Coercion, prejudice, and racism were all central parts of the fabric of life and of the experience of black people, both slaves and free.

Notes

1 R. Matthee, "Exotic Substances: The Introduction and Global Spread of Tobacco, Coffee, Cocoa, Tea, and Distilled Liquor, Sixteenth to Eighteenth Centuries," in *Drugs and Narcotics in History*, eds. R. Porter and M. Teich (New York, 1995), 38–46; J. Walvin, *Fruits of Empire: Exotic Produce and British Taste, 1660–1800* (London, 1997).

2 R.J. Ferry, "Encomienda, African Slavery, and Agriculture in Seventeenth-Century Caracas," *Hispanic American Historical Review*, 61 (1981), 620–36.

3 A.L. Butler, "Europe's Indian Nectar: The Transatlantic Cacao and Chocolate Trade in the Seventeenth Century" (M. Litt., Oxford, 1993).

4 J.R. Ward, "The Profitability of Sugar Planting in the British West Indies, 1650–1834," *Economic History Review*, 2nd ser., 31 (1978), 208.

5 A.E. Smith, *Colonists in Bondage: White Servitude and Convict Labor in America, 1607–1776* (Chapel Hill, NC, 1947).

6 D. Hancock, ed., *The Letters of William Freeman, London Merchant, 1678–1685* (London, 2002), xl.

7 C.R. Boxer, *Salvador da Sá and the Struggle for Brazil and Angola 1602–1686* (London, 1952) and *The Dutch in Brazil, 1624–1654* (Oxford, 1957).

8 D.H. Akenson, *If the Irish Ran the World: Montserrat, 1630–1730* (Montreal, 1997).

9 H. Beckles, "The Economic Origins of Black Slavery in the British West Indies, 1640–1680: A Tentative Analysis of the Barbados Model," *Journal of Caribbean History*, 16 (1982), 36–56, esp. 52–4.

10 H. Beckles, *White Slavery and Black Servitude in Barbados, 1627–1715* (Knoxville, TN, 1989); L. Gragg, "'To Procure Negroes': The English Slave Trade to Barbados, 1627–60," *Slavery and Abolition*, 16 (1995), 70, 74.

11 L.H. Roper, *Advancing Empire: English Interests and Overseas Expansion, 1613-1688* (Cambridge, 2017); J. Morgan, *Reckoning with Slavery: Gender, Kinship, and Capitalism in the Early Black Atlantic* (Durham, NC, 2021).

12 H. Beckles and A. Downes, "The Economics of Transition to the Black Labor System in Barbados, 1630–1680," *Journal of Interdisciplinary History*, 18 (1987), 246–7.

13 N. Zahedieh, "Trade, Plunder, and Economic Development in Early English Jamaica, 1655–89," *EcHR*, 2nd ser., 39 (1980), 205–22.

14 V.A. Shepherd, "Livestock and Sugar: Aspects of Jamaica's Agricultural Development from the Late Seventeenth to the Early Nineteenth Century," *Historical Journal*, 34 (1991), 627–43.

15 T. Burnard, "Who Bought Slaves in Early America? Purchasers of Slaves from the Royal African Company in Jamaica, 1674–1708," *Slavery and Abolition*, 17 (1996), 88.

16 R. Law, "The First Scottish Guinea Company, 1634–9," *Scottish Historical Review*, 76 (1997), 185–202.

17 A.L. Araujo, *Slavery in the Age of Memory: Engaging the Past* (London, 2021).

18 D. Geggus, "The French Slave Trade: An Overview," *William and Mary Quarterly*, 3rd ser., 58 (2001), 120.

19 A.M. Carlos and J.B. Kruse, "The Decline of the Royal African Company: Fringe Firms and the Role of the Charter," *EcHR*, 2nd ser., 49 (1996), 291–313; K. Morgan, ed., *The British Transatlantic Slave Trade. II. The Royal African Company* (London, 2003).

20 F. Guerra, "The Influence of Disease on Race, Logistics and Colonization in the Antilles," *Journal of Tropical Medicine and Hygiene*, 69 (1966), 33–5.

21 T. Burnard, "'The Countrie Continues Sicklie': White Mortality in Jamaica, 1655–1780," *Social History of Medicine*, 12 (1999), 45–72, esp. 55–6, 71.
22 E.S. Morgan, "The First American Boom: Virginia 1618 to 1630," *William and Mary Quarterly*, 3rd ser., 28 (1971), 169–98, esp. 197.
23 R. Menard, "From Servants to Slaves: The Transformation of the Chesapeake Labor System," *Southern Studies*, 16 (1977), 355–90, esp. 389; J. Horn, *Adapting to a New World: English Society in the Seventeenth-Century Chesapeake* (Chapel Hill, NC, 1994).
24 A.T. Vaughan, "The Origins Debate: Slavery and Racism in Seventeenth-Century Virginia," *Virginia Magazine of History and Biography*, 97 (1989), 353.
25 R.C. Nash, "South Carolina and the Atlantic Economy in the Late Seventeenth and Eighteenth Centuries," *EcHR*, 2nd ser., 45 (1992), 677–702.
26 E.W. Evans and D. Richardson, "Hunting for Rents: The Economics of Slaving in Pre-Colonial Africa," *EcHR*, 2nd ser., 48 (1995), 665–86.
27 With reference to the port of Annamaboe and the Fante, R. Sparks, *Where the Negroes Are Masters: An African Port in the Era of the Slave Trade* (Cambridge, MA, 2014).
28 P. Manning, *Slavery and African Life: Occidental, Oriental and African Slave Trades* (Cambridge, 1990).
29 R. Shumway, *The Fante and the Transatlantic Slave Trade* (Rochester, NY, 2011).
30 J. Thornton, *Africa and the Africans in the Making of the Atlantic World, 1400–1800* (2nd ed., Cambridge, 1998), 98–125 and *Warfare in Atlantic Africa, 1500–1800* (London, 1999), esp. 127–39.
31 R. Law, "'Here Is No Resisting the Country': The Realities of Power in Afro-European Relations on the West African 'Slave Coast'," *Itinerario* 18 (1994), 50–64, esp. 52–7; W. St. Clair, *The Grand Slave Emporium: Cape Coast Castle and the British Slave Trade* (London, 2006).
32 G. Nováky, *Handelskompanier och kompanihandel: Svenska Afrikakompaniet 1649–1663* (Uppsala, 1990), English summary, 241, 244.
33 R. Law and K. Mann, "West Africa in the Atlantic Community: The Case of the Slave Coast," *William and Mary Quarterly*, 3rd ser., 56 (1999), pp. 307–34; T. Green (ed.), *Brokers of Change: Atlantic Commerce and Cultures in Pre-Colonial Western Africa* (Oxford, 2012).
34 R.B. Allen, *European Slave Tradign in the Indian Ocean, 1500-1850* (Athens, OH., 2015).
35 K.Y. Daaku, *Trade and Politics on the Gold Coast, 1600–1700* (Oxford, 1970), 96–114.
36 G.U. Nwokeji, *The Slave Trade and Culture in the Bight of Biafra: An African Society in the Atlantic World* (New York, 2010).
37 J. Thornton, "Warfare, Slave Trading and European Influence: Atlantic Africa 1450–1800," in *War in the Early Modern World*, ed. J. Black (London, 1999), 141.
38 E. van den Boogaart and P.C. Emmer, "The Dutch Participation in the Atlantic Slave Trade, 1596–1650," in *The Uncommon Market: Essays in the Economic History of the Atlantic Slave Trade*, ed. H.A. Gemery and J.S. Hogendorn (New York, 1979), 367; K.G. Davies, *The Royal African Company* (London, 1957), 292.
39 J.C. Miller, "Some Aspects of the Commercial Organization of Slaving at Luanda, Angola, 1760–1830," in *Uncommon Market*, ed. Gemery and Hogendor, 104.
40 G. O'Malley, *Final Passages: The Intercolonial Slave Trade of British America, 1619-1807* (Chapel Hill, NC, 2014).
41 K. Gerbner, *Christian Slavery: Conversion and Race in the Protestant Atlantic World* (Philadelphia, PA, 2018).

42 S.E. Smallwood, *Saltwater Slavery: A Middle Passage from Africa to American Diaspora* (Cambridge, MA, 2007).

43 C.C. Bailey, *African Voices of the Atlantic Slave Trade: Beyond the Silence and the Shame* (Boston, MA, 2005); V. Carretta, *Equiano the African: Biography of a Self-Made Man* (Athens, GA, 2005).

44 W. Hawthorne, "'Being Now, as it Were, One Family': Shipmate Bonding on the Slave Vessel *Emilia*, in Rio de Janeiro and throughout the Atlantic World," *Luso-Brazilian Review*, 45 (2008), 53–77.

45 A. Diptee, *From Africa to Jamaica: The Making of an Atlantic Slave Society, 1775–1807* (Gainesville, FL, 2010); J.H. Sweet, "Defying Social Death: The Multiple Configurations of the African Slave Family in the Atlantic World," *William and Mary Quarterly*, 3rd ser., 70 (2013), 251–72.

46 K.M. de Queirós Mattoso, *To Be a Slave in Brazil* (New Brunswick, NJ, 1989); M.L. Conniff and T.J. Davis, *Africans in the Americas: A History of the Black Diaspora* (New York, 1994); C.H. Lutz, *Santiago de Guatemala, 1541–1773: City, Caste, and the Colonial Experience* (Norman, OK, 1994); J. Landers, *Black Society in Spanish Florida* (Urbana, IL, 1999); H.L. Bennett, *Africans in Colonial Mexico: Absolutism, Christianity, and Afro-Creole Consciousness, 1570–1640* (Bloomington, IN, 2003).

47 S. White, "Slaves and Poor Whites: Informal Economies in an Atlantic Context," in *Louisiana: Crossroads of the Atlantic World*, ed. C. Vidal (Philadelphia, PA, 2014), 102.

48 J. Carney, *Black Rice: The African Origins of Rice Cultivation in the Americas* (Cambridge, MA, 2001).

49 A. Sluyter, *Black Ranching Frontiers: African Cattle Herders of the Atlantic World, 1500-1900* (New Haven, CT, 2012); K. Dawson, *Undercurrents of Power: Aquatic Culture in the African Diaspora* (Philadelphia, PA, 2018).

50 D. Lambert, *Mastering the Niger: James MacQueen's African Geography and the Struggle over Atlantic Slavery* (Chicago, Ill., 2013).

51 P.M.S. Silva, *Urban Slavery in Colonial Mexico: Puebla de los Angeles, 1531-1706* (Cambridge, 2018).

52 J.H. Sweet, *Recreating Africa: Culture, Kinship, and Religion in the African-Portuguese World, 1441–1770* (Chapel Hill, NC, 2003).

53 A.J.R. Russell-Wood, "Black and Mulatto Brotherhoods in Colonial Brazil: A Study in Collective Behaviour," *Hispanic American Historical Review*, 54 (1974), 567–602, esp. 581–2; R.C. Rath, "African Music in Seventeenth-Century Jamaica: Cultural Transit and Transmission," *William and Mary Quarterly*, 3rd ser., 50 (1993), 700–26.

54 T. Nunley, *At the Threshold of Liberty: Women, Slavery, and Shifting Identities in Washington, DC* (Chapel Hill, NC, 2021).

55 P. Manning, *The African Diaspora: A History through Culture* (New York, 2009).

56 L.M. Rupert, *Creolization and Contraband: Curaçao in the Early Modern Atlantic World* (Athens, GA, 2012), 242.

57 W. Hawthorne, *From Africa to Brazil: Culture, Identity, and an Atlantic Slave Trade, 1600–1830* (New York, 2010).

4 The Slave Trade at its Height

Late 1781 is known mostly for the British surrender at Yorktown, but, the following month, the *Zong* massacre threw appalling light on the cruelty of the slave trade. A Dutch slave ship based in Middelburg in the Netherlands was captured with slaves on board by the British in February 1781 and sold the following month to a syndicate of Liverpool merchants. Sailing from Accra with 442 slaves on 18 August, the *Zong* had far more slaves than a British ship of that size would carry. The *Zong* reached the Caribbean but was affected by disease, short of water, and poorly commanded. From 29 November, 142 slaves were killed by being thrown overboard. Their value was then claimed on the ship's insurance which led to a legal controversy in 1783 over the insurer's liability, which was finally rejected by the courts. The episode was a longstanding cause célèbre for Abolitionists and J.M.W. Turner produced a dramatic painting of the episode entitled *The Slave Ship* in 1840. It has also played a role more recently, as in Fred D'Aguiar's novel *Feeding the Ghosts* (1997) and the film *Belle* (2014).

In his novel *L'An 2440* (1770), written while the slave trade was reviving after the Seven Years' War, the radical French writer Louis Sébastien Mercier (1740–1814) described a monument in Paris. This depicted a black man, his arms extended, rather than in chains, and a proud look in his eye. He was surrounded by the pieces of 20 broken scepters and atop a pedestal with the inscription, *Au vengeur du nouveau monde* (To the avenger of the New World). To his readers, this would have seemed a utopian prospect and also a proof of Mercier's radicalism. The idea was radical because of the depiction of triumphant, free, one-time slaves. It was also radical because of the interpretation thus given to the stadial theory of change, which discussed the development of human society in terms of stages. This was an influential concept in Western thought, but was generally presented in terms of progress as leading to the triumph of Western civilization. The idea that progress would lead to a world in which monarchies had fallen victim to free slaves was truly subversive.

DOI: 10.4324/9781003457923-4

At the same time, Mercier's image captured anew the continual relationship between slavery, the slave trade, and wider currents of development. In the case of Mercier, the last was the intellectual movement in the West subsequently known as the Enlightenment. This was a tendency, not a movement; a tendency for the progressive application of reason in the cause of critical inquiry and improvement. This tendency made many existing practices appear in need of change, if not radical transformation. In particular, there was an attempt to emulate the Scientific Revolution of the late seventeenth and early eighteenth centuries with a science of man. In the case of some commentators, a rethinking of the nature of humanity came to pose a challenge to slavery and the slave trade, indeed a fundamental challenge. At the same time, as a reminder of the complex impact of change, other trends in the century—notably, economic development—furthered the expansion of the slave trade. Indeed, this was the century in which the Atlantic slave trade came to its height. Moreover, any analysis of the growth late in the century of Western opposition to the slave trade has to comprehend not only Enlightenment attitudes but also another stage in the long debate about slavery among Christian thinkers.

Offering an alternative to the Bible, literature was to be of major significance to the changing public attitudes in the West that affected the slave trade by the close of the century. Books were important for a number of reasons. In the absence of photography, books certainly brought home at least part of the nature of slavery to European readers. This was true of both factual works and novels. In his highly successful novel *Candide* (1759), the leading Enlightenment French writer François Marie Arouet de Voltaire (1694–1778) had his protagonist visit Surinam (on the Atlantic coast of South America), which had been colonized by the Dutch as a plantation economy in the seventeenth century. Voltaire himself never left Europe. A black man told Candide:

> Those of us who work in the factories and happen to catch a finger in the grindstone have a hand chopped off; if we try to escape, they cut off one leg. Both accidents happened to me. That's the price of your eating sugar in Europe ... Dogs, monkeys, and parrots are much less miserable than we are. The Dutch ... who converted me, tell me every Sunday, that we are all children of Adam.[1]

The last sentence is a reference to Christian hypocrisy directed at the Calvinist Dutch. In 1804, 45 years after the appearance of *Candide*, the National Convention abolished slavery in all French colonies. In the radical crucible of the French Revolution, Enlightenment horror had become state policy, as a politicized science of man was legislated into action.

Yet, many Western writers did not share radical or even Enlightenment ideas. Thus, the British writer Tobias Smollett (1721–71), who went to the Caribbean in 1740–1, would be very ripe for "cancellation" in the cultural politics of the present. Part of his prosperity derived from the wealth of his wife Anne, an heiress whose father had bought into Jamaican plantations.

To modern eyes, there is a harshly matter-of-fact tone in Smollett's correspondence when pressing for profit from the plantations, not least through selling slaves; and, as was unfortunately the norm, the latter is discussed without considering the personal links that would be thereby ruptured.

This matter-of-fact tone is also found in his writing. Thus, in *Roderick Random* (1748), the novel in which there are most references to slavery, Thomson, "utterly unable to subsist any longer" in London, offers "his service in quality of mate to the surgeon of a merchant's ship bound to Guinea [West Africa], on the slaving trade," only to gain a naval position. There is no expression of shock. So also with "a surgeon and overseer to his plantations" for a Jamaican gentleman, who is able to order "a couple of stout negroes" to escort a guest, and to provide Random with plentiful food and drink; without, of course, any reference to the far more modest diet of the slaves. Later in the book, Random joins a slaver that buys 400 slaves in Africa, selling in Spanish-ruled Buenos Aires those that survive fever in the passage. This is a particularly long and therefore more deadly passage across the South Atlantic. Smollett refers to "the disagreeable lading of negroes,"[2] but does not suggest that that is due to any sympathy for the slaves. They are treated by Smollett neither as individuals nor as a group deserving of sympathy.

In Smollett's *Peregrine Pickle*, a discussion about whether cats can be eaten includes the observation "that the Negroes on the coast of Guinea, who are a healthy and vigorous people, prefer cats and dogs to all other fare,"[3] an assessment that was presumably intended to arouse a degree of contempt. A reference to Africans in *Travels through France and Italy* may relate to slaves, and is certainly one in which they are worse treated. Referring to French fish that is spoiled, Smollett added: "At best it must be in such a mortified condition, that no other people, except the negroes on the coast of Guinea, would feed upon it."[4] Smollett, in the following letter in his *Travels*, emphasized the relative, culturally dependent, nature of acceptability, but that is not the point here.

In Smollett's writing, there is no ready identification of slaves with sub-Saharan Africans. Indeed, in his criticism of classical Rome, Smollett observes: "The execrable custom of sacrificing captives or slaves at the tombs of their masters and great men, which is still preserved among the negroes of Africa, obtained also among the ancients, Greeks as well as Romans." Smollett had no direct knowledge about African circumstances.

Separately, visiting the Sardinian galleys at Villefranche near Nice, Smollett saw the slaves who rowed them:

> miserable wretches, chained to the banks … This is a sight which a British subject, sensible of the blessings he enjoys, cannot behold without horror and compassion. Not but that if we consider the nature of the case, with coolness and deliberation, we must acknowledge the justice, and even sagacity, of employing for the service of the public, those malefactors who have forfeited their title to the privileges of the community.

Referring to these as slaves, Smollett, however, clearly differentiates them from those who are not criminals:

> It is a great pity, however, and a manifest outrage against the law of nations, as well as of humanity, to mix with those banditti, the Moorish and Turkish prisoners who are taken in the prosecution of open war. It is certainly no justification of this barbarous practice, that the Christian prisoners are treated as cruelly as Tunis and Algiers.[5]

Smollett makes clear the grim nature of life onboard the gallies (of which the British had none), with the rowers swarming with vermin, in other words lice. In his novel *Peregrine Pickle*, Cadwallader becomes a French galley slave as a result of insulting the king. As well as mentioning Christian slaves held by Christian powers, there are Christian slaves held by non-Christian powers. Thus, Major Macleaver had been a slave in Algiers, the leading center of slave-raiding against Christian Europe. This was an accurate account of the situation, rather than one that reflected any attempt to contextualize, extenuate, or diminish, slavery by Europeans in the Atlantic World.

The response of novelists to slavery was not generally that of modern commentators. Thus, in Henry Mackenzie's sentimental novel *Julia de Roubigné* (1777), a tale of tragic love, the disappointed lover determines not to whip his slaves, which is a long way from freeing them. Moreover, unlike with the Dutch colony of Surinam in Voltaire's *Candide* (1759), a work that does not hide the terrible cruelty shown to slaves and the hypocrisy of the slave-owners, there is usually no indication of the extent to which goods that are mentioned have been obtained through the use of slave labor. In *Roderick Random*, Dr Wagtail praises coffee "which in cold phlegmatic constitutions like his, dried up the superfluous moisture, and braced the relaxed nerves," being "utterly unknown to the ancients" (45) and thus the product of more recent European expansion. Such comparisons with the ancients was designed to show the superiority of the modern world, as part of the Ancients versus Moderns debate. There is no mention of the extent to which coffee was produced in Caribbean slave economies.[6]

Voltaire's point about Christian hypocrisy was to be echoed in 2003 when, on a visit to Africa, the American President, George W. Bush, traveled to the major West African slave-trading post at Gorée, which had been used by the Dutch, French, and British. Its value as a slave port meant that Gorée had been fought over, as well as being a prime topic in peace negotiations, held to determine the post-war allocation of transoceanic territories, as in 1762 and 1782 when the 1763 and 1783 treaties between Britain and France were negotiated. Now in Senegal, near the modern capital, Dakar, Gorée was a choice of destination designed to send a message about Bush's concern for African-Americans, as well as his awareness of their distinctive history, and his grasp of the role of suffering to it. Echoing an evangelical theme that became important from the late eighteenth century, Bush declared that "Christian men and women became blind to the clearest commands of their faith … Enslaved Africans discovered a suffering Savior and found him as more like themselves than their masters." President Obama visited Gorée in 2013.

Gorée and other ports were indeed very busy in the eighteenth century. It was the peak period of the slave trade, with about half of those shipped from Africa to the Americas in the period 1450–1900 moved in this century alone.[7]

There was a clear sense of the trade becoming larger in scale, more sophisticated, and greater in importance. The mechanics of the trade require consideration before we discuss the criticism that was to undermine it and finally lead to its end.

Selling to Others

As before, the Atlantic slave trade involved selling slaves to the colonies of other powers, and also to one's own colonies. Most obviously, both France and Britain sought to profit from demand in Latin America and from the wealth of its economies. In doing so, they contributed greatly to the many and significant illicit slave flows, flows that challenge the quantification of the slave trade on which so much excellent work has been done.

Nevertheless, in order to avoid the problems posed by the Spanish regulatory regime, which covered so much of the Americas from the Spanish colony of Florida southwards, it was far more desirable to gain permission to trade with Spanish America. Indeed, in 1701, as a sign of closer Franco-Spanish relations following the accession of the Bourbon Philip V (younger grandson of Louis XIV) to the Spanish throne the previous year, the French Guinea Company was granted the *Asiento* contract to transport slaves to Spanish America for ten years. This was a lucrative opening into the protected trade of the Spanish Empire, and one that was highly unwelcome to the British.

In turn, the victorious British gained the *Asiento* and the right to trade with Spanish America in 1713 at the close of the War of the Spanish Succession (1702–13), a conflict in which they had defeated the French. This right, and, as with free trade, the word is bitterly inappropriate, was exercised by the South Sea Company. The extravagant commercial hopes the company gave rise to helped launch the vast speculative bubble in its shares, a bubble that reflected hopes about trade to the Americas and the Pacific. The share value crashed in 1720, causing a major political scandal about governmental connivance in fraudulent financial practices. Speculation in shares was an aspect of the public world of finance and politics that became of growing significance in the West in the eighteenth century. This world brought together various developments that have interested historians, including what they have termed the Financial Revolution, the Commercial Revolution, and the development of public space. London and Amsterdam were key centers of this activity. This was a world of stock markets and of financial reporting in newspapers. There was a link to "political arithmetic"—the attempt to develop and apply statistics. In Britain, this was related to the growth of parliamentary government after the Glorious Revolution of 1688–9 and to Parliament's role in commercial regulation.[8]

This was the active context within which opportunities were gauged and pursued, one in which business was linked to politics. The discussion of how best to develop the slave trade was an aspect of this consideration, and, centrally, how best to regulate it. This serves as a reminder that, under scrutiny, the idea of clear national interests dissolves. Instead, both policy and its implementation were debated and contested. For example, the ambitious role of the South Sea Company created tensions in the British slave trade, with, in 1724, the Royal African Company pressing the ministry for assistance against the company. The eventual pressures for the regulation, then limitation, and, finally, abolition of the slave trade, were aspects of this debate over policy and its implementation.

Unlike Britain and France, powers that lacked important colonies were largely dependent on selling to the colonies of other states. The Dutch had bases on the Gold Coast of West Africa, including Axim, Hollandia, Accadia, Butri, and Shama, but in the Americas lacked a market comparable to Portuguese Brazil, French Saint-Domingue (modern Haiti), or British Jamaica, as well as the other French and British colonies. Instead, the Dutch sold to all they could reach through entrepôts on their West Indies' islands of Curaçao and St. Eustatius, and carried about 310,000 slaves during the century.[9] Based on Cape Town, which had been founded in 1652 as a base on the way to the Indian Ocean, Cape Colony was another Dutch slave society.[10]

In a weaker position, the Brandenburg (Prussian) and Danish companies were unable to make money this way. Indeed, in 1717, the two forts that the Brandenburg Company had on the Gold Coast, Fort Dorothea (Accadia) and Fort Friedrichsburg (Hollandia), were sold to the Dutch, whose purchase of them reflected a sense of profitable opportunity.

In succession, three Danish West Indian companies failed to make the necessary profits; the Danes owned several small islands in the West Indies—St. Croix, St. John, and St. Thomas (sold in 1917 to the USA, and now the American Virgin Islands)—but lacked a large market. The Danish colonial presence on the African Gold Coast had begun in the 1650s, with Christiansborg acquired from Sweden in 1653 and Fort Augustenborg in 1700. The motives of the Danish government, which supported the companies involved financially, and with monopoly rights and other privileges, were mercantilist. The intention was that Denmark should obtain part of the great wealth created by the international trade in slaves and sugar in order to strengthen key Danish commercial groups, and thus the country economically and financially. In a narrower sense, there was also a drive to gain larger revenues for the Danish Treasury. Thus, entry into the slave trade can be seen as an aspect of a more general government policy that was intended to further maritime trade and industry, and strengthen the state. The Danish sugar industry duly grew, notably from the 1750s. Expanding in the 1780s, partly due to war between Britain and the Dutch in 1780–4, so that Denmark benefited from neutrality, the Danish slave trade peaked in 1787.[11]

This expansion was an instance of the way in which the slave trade, both in its general trend and in the specifics of market shares, was greatly affected by war.

The Danish role in the slave trade really started in the eighteenth century. So did that of Sweden, after the short-lived attempt in the mid-seventeenth century. Gustavus III of Sweden, a monarch usually classed as one of the Enlightened despots, gained the Caribbean island of St. Barthélemy from France in 1784; it was returned in 1877. In return for the island being ceded to Sweden, Gustavus granted his ally France a depot for naval stores at Gothenburg in Sweden. The ambitious Gustavus had ideas about Swedish colonies in Africa, but the person he sent out to gather information, Carl Bernhard Wadström, instead became a rather prominent Abolitionist.

More generally, the slave trade was not a constant, neither at the national level nor at the aggregate one. Flows varied, as the sources and destinations of slaves changed. The majority of Africans transported in the eighteenth century went to the West Indies and Brazil, and fewer than a fifth to Spanish or North America. The biggest shippers, in order, were the British, Portuguese, French, and Dutch. Anglophone scholarship concentrates heavily on Britain, but France also was a key supplier to the West Indies and it is pertinent to note its role.

The French Trade

During the eighteenth century, the French colonies obtained 1,015,000 slaves from French sources, although an illegal British trade to these colonies was important, which makes it difficult to assess the numbers imported into them.[12] In 1788, the French West Indies contained 594,000 slaves, many harshly treated. In 1687, Saint-Domingue, the largest French colony in the West Indies (modern Haiti), contained 4,500 white people and 3,500 black people; in 1789, 28,000 white people, 30,000 free black people, and 406,000 slaves. In the 1780s, thanks to strong reexports into Europe of sugar imported into France, sugar prices rose despite a general recession in France.[13]

With slavery closely linked to profit, this was the peak decade for the receipt of slaves by the French West Indian colonies—nearly 30,000 annually. Indeed, the numbers sent to Saint-Domingue rose from 14,000 annually in 1766–71 to 28,000 annually in 1785–9, and the strength of Saint-Domingue's economy ensured that it was able to get a better choice of slaves. Large numbers of slaves were also sent to the French West Indian islands of Guadeloupe and Martinique. The French, having established their first base in Louisiana in 1699, imported the first slaves to the colony in 1719, although it never became a major plantation economy, and thus an important slave society. The attempt to develop large-scale tobacco production there was unsuccessful. There were generally about 2,000 African slaves in Louisiana, many used for household tasks. The major French source of slaves was the basin of the River Senegal, in modern Senegal, via the slaving ports of Gorée and St. Louis. In contrast, Assinie on the Ivory Coast was only held by the French from 1687 to 1705, while Forcados on the Benin coast was only held from 1786 to 1792, a period when the French slave trade was responding to the greatly expanded demand for slaves in Saint-Domingue.

In a separate trade, slaves obtained from East Africa and Madagascar supplied the French colonies in the Indian Ocean. These colonies were out-liers of the Atlantic world that were crucial to the geopolitics of French power in the Indian Ocean, supporting French activity in India and further east. The island of Réunion, which they had claimed in 1642, was a source of coffee. The formerly Dutch-held island of Mauritius, acquired by France in 1715, was a source of sugar. In 1769–72, French expeditions acquired clove plants on Ambon in the East Indies and introduced them to Mauritius. Such activity linked knowledge, power, determination, and the global range of Western activity. The slave economy was both a central part of this range and a key product of it. People, like goods and plants, were treated as commodities.

Plantation Goods

As a result of the efforts of the slaves, exports from the Americas boomed, helping to lead to a major rise in European consumption: of coffee in Europe across the century from 2 million to 120 million pounds, of chocolate from 2 million to 13 million pounds, and of tobacco, especially from Virginia, from

50 million to 125 million pounds. In each case, this was a rise far above that of the European population. It was possible to produce some goods in Europe, including tobacco and sugar, but for a number of reasons, principally climate, production there was more precarious and could not match that in the Americas. The arrival in Europe of fleets bringing plantation goods was watched carefully as they greatly affected the availability and price of the goods they brought.

This oceanic trade greatly influenced that of Europe. For example, the French West Indian islands were particularly important for the production of sugar. This production helped drive forward French trade and production, interacting with opportunities in French domestic and foreign markets. At the beginning of the century, Bordeaux's sugar refiners enjoyed the privilege of transporting their product to much of France without paying many of the internal tolls of which their rivals in La Rochelle, Marseilles, and Nantes complained. Partly as a result, in the recovery that followed the end of war with Britain in 1713, Bordeaux's imports of sugar, indigo, and cocoa from the French West Indies tripled in 1717–20. This was the beginning of a massive increase in French reexports to northern Europe. These exports competed directly with those of Britain in key markets such as Hamburg. Production rose during the century. In 1778, Saint-Domingue exported 1,634,032 quintaux of sugar (100 kilograms to a quintal).

Sugar was not alone. After the Caracas Company was founded in 1728 and given monopoly rights by the Spanish government to transport cacao from Venezuela, exports of cacao to Europe increased, and the already established growth in production there continued, which encouraged a major rise in slave imports. Until 1784, the company enforced its monopoly by patrolling the coast as part of a campaign against smugglers. The Caracas Company, like the other chartered companies, reflected an important aspect of early modern Western government—namely, the extent to which its functions were "privatized," or, rather, handled by other regulatory bodies. At the same time, the slave trade also interacted with aspects of a modernizing West. Two in particular were notable: the role of capitalism and the central place of the state in the shape of a powerful navy. The first was seen both in the development of the slave trade and in that of the plantation economies. For example, by the 1790s, in response to growing demand in Europe and the economies of scale, the cacao plantations in Venezuela were concentrated into fewer hands, with the holdings and slave forces of each growing in scale: eight families controlled production.[14]

Coffee was another major product from the New World and, thanks to a range of factors in which slavery was integral and crucial, the Western economies took over the bulk of the world trade in coffee, most notably replacing Yemen as the major source for the coffee drunk in the Ottoman Empire (Turkey). The coffee house, the West's principal site for urbanity, the leading "public space" of sociability, rested ultimately on slave production in the West Indies. Aside from the prominent goods, notably sugar, coffee, and cocoa, the provision of other New World products depended on slaves. Thus, the British logwood cutters who settled in what is now Belize in Central America brought in African slaves from 1720. From the 1760s, the British brought in more slaves to help with mahogany cutting, which required more labor. These were largely male, as women were not used for logging: the work and the environment were hard. This development ties in with the increased fashion of using mahogany in Western furniture building.

The role of the New World in the Western economic system continued strong because of the failure to produce significant and competitive quantities of its tropical goods elsewhere. Whereas New England and North Carolina iron and timber goods competed in the British market with Scandinavian production of both as well as with British production, there was no equivalent competition for sugar. With sugar, India was not to take a role competing with New World production, as it did with cotton and cotton clothes. In 1792, the West Indies' interest in Britain lobbied hard and successfully against the attempt by the British East India Company to export sugar from India, which might have led to a cut in the price of sugar.[15] Nor yet was there significant sugar-beet production in Europe.

The impact of trade with the New World rippled out from Europe's Atlantic ports. In part, this reflected the activity of merchant networks, which were not those of the transoceanic trading groups but which profited from their imports of colonial goods. The reexport of these colonial goods came to play a major role in the European economy.

Conflicts over the Atlantic Trade

A series of wars shook the Atlantic world in the eighteenth century, wars between the Western oceanic powers that greatly affected the slave trade. In 1739, for example, Anglo-Spanish differences over trade in the Caribbean led to the War of Jenkins' Ear, named after the ear of the captain of a British merchantman cut off by Spanish officials seeking to prevent illicit trade. Much of the commerce in question was linked, at least indirectly, to the slave trade, with Britain trying to gain access to the wealth of the Spanish New World economy. In part, this attempt involved legal trade. There was indeed a considerable degree of interdependence between the British and the Spanish New World economy, not least the extension of credit to the Spanish-American buyers of British exports. However, a growing reluctance on the part of the British to accept Spanish terms for legal trade and inter-dependence interacted with Britain's determination to develop its maritime empire. The result, in 1739, was war.[16] In 1739–41, the conflict with Spain focused on the control of Caribbean trade, with British expeditions directed against ports held by Spain, notably Porto Bello in 1739 and, with far less success, Cartagena in modern Colombia in 1741.

Those pressing in Britain for an assertive policy in the West Indies in the 1730s were able to identify themselves with a strong image of the national interest, an image in which trade, maritime destiny, and overseas activity all aligned. Slavery did not yet compromise this image, as it was to do in the 1800s. In 1757, when Britain and France were at war, a pamphlet calling for an expansionist policy in North America and the West Indies added West Africa:

> The vitals of our West Indian islands are our African settlements … though we have been possessed of the trade, particularly on the Gold Coast, Whydah, and Gambia, upwards of eighty years, the French are daily undermining us there, so that if by open force they do not exclude us from all trade to Africa, they will at least by degrees worm us out of it, as they have already done upon the Gum Coast; if we do not immediately take such salutary measures, as may effectually frustrate this long and deep laid design of the French.[17]

Once war had begun, the sugar islands were a major site of conflict. France's production of colonial goods and its slave trade were affected by repeated war with Britain and eventually, in the 1790s, by a major and, in the end, successful slave rebellion on Saint-Domingue. This slave rebellion was linked both to the French Revolution and to the war with Britain which broke out in 1793. This was the last that century of a sequence of wars between Britain and France that involved conflict in West Africa and the Americas. These conflicts affected the slave trade and the trade from the New World which it fed.

In the War of the Austrian Succession, in which the British fought France in 1743–8, they mounted few transoceanic operations, although they captured Port Louis on Saint-Domingue in 1748. More significant was the attack on French overseas trade, notably with the French West Indies, an attack that greatly affected the exports from the slave plantations. As a result, by 1748, the price of goods from the French West Indies was very high in European entrepôts such as Hamburg.[18] British pressure on the trade of France's colonies affected the funds available there for the purchase of slaves. After the war ended, large quantities of sugar and coffee from the French West Indies appeared in Hamburg,[19] causing their price to fall.[20] Britain and France quarreled in 1749 over the possession of the Caribbean island of Tobago, but their vexed relations in the early 1750s focused on North America, leading to conflict there in the River Ohio country in 1754 and to the declaration of war in 1756. In the Seven Years'

War (1756–63), however, the British devoted a much greater effort to the seizure of the sugar islands than in the previous war. From 1758, there was a systematic attempt to capture the French colonies. Guadeloupe was captured in 1759, and Grenada, Martinique, St. Lucia, and St. Vincent in 1762. The French slave stations in Senegal in West Africa—St. Louis and Gorée—were taken in 1758. Havana was captured from Spain in 1762. Under the peace settlement, the Peace of Paris, of 1763, Martinique, Guadeloupe, Cuba, Gorée, and St. Lucia were returned by victorious Britain, but Grenada, St. Vincent, and St. Louis in Senegal were retained. The slave trade played a significant role in the negotiations. John, 3rd Earl of Bute, George III's key adviser, had noted "When Choiseul [the French foreign minister] gives Senegal up and seems facile on Gorée, it is with an express proviso that the French be put in possession of a sea port on the Slave Coast."[21] In the event, Britain sought, but without success, to develop Senegambia as a crown colony that would include plantations worked by slaves as well as slave soldiers.[22]

In the War of American Independence, in which France came into the war in support of the American revolutionaries in 1778, control over the slave stations in Senegal changed hands in 1779. In the West Indies, the French took Dominica (1778), Grenada (1779), St. Vincent (1779), Tobago (1781), Nevis (1782), St. Christopher (1782), and Montserrat (1782), although the British captured St. Lucia (1778). The importance of the slave plantations to the British economy helped explain the focus of British military operations in North America on Georgia, the Carolinas, and Virginia from 1778. They were regarded as a key source of supplies for the British West Indies, and especially of food, without which the viability of the slave economy was limited. The British slave trade was hit by the war, alongside the profitability of the sugar plantations.[23] In the eventual peace settlement, the Treaty of Versailles of 1783, France gained Tobago and Senegal, a slave island and key slave trade bases. The two were linked.

In the French Revolutionary War, which broke out in 1793, the slave trade and the plantation economies of the West Indies were disrupted anew. The British captured Gorée in 1800, and, in the West Indies, Tobago in 1793, St. Lucia in 1794 and (after it had been retaken in 1795) in 1796, Trinidad in 1797, Surinam in 1799, Curaçao in 1800, and the Danish and Swedish West Indian islands in 1801. Alongside British naval primacy and the slave rebellion on Saint-Domingue, these captures inflicted serious damage on the French slave economy, as well as on that of its allies. This damage undermined the transoceanic dimension of the French economy and encouraged its focus on the European dimension. Moreover, this damage weakened the significance of the West Indies to France and its allies, while increasing the value of the West Indies to a Britain that was increasingly excluded, by French successes, from Europe's trade.

The Portuguese Slave Economy

In contrast to the impact of Britain and rebel slaves on the French slave economy, the Portuguese slave trade in the South Atlantic did not, in the eighteenth century, face a challenge comparable to that mounted by the Dutch in the seventeenth century to its colonies of Angola and Brazil. Angola, where the major bases were Luanda and Beneguela, supplied about 2 million slaves during the eighteenth century, mostly to Brazil. The Portuguese had more bases further north, especially in Portuguese Guinea (Guinea-Bissau) at Cacheu. Like the other European slaving nations, the Portuguese also traded for slaves along coasts where they did not have any bases. Off the Atlantic coast, the Portuguese had the islands of Annobon, Fernando Póo, Principe, and São Tomé, although Annobon and Fernando Póo were gained by Spain in 1776; otherwise, Spain had no direct presence in the African slaving world.

Growing demand for slaves in Brazil reflected its major economic expansion. Brazil was one of the economic successes of the eighteenth century, and thus an indication, like the major growth of cotton production in the American South in the nineteenth century, that slavery was compatible with economic progress—indeed, a central aspect of it: both contributory factor and consequence. At the same time, alongside the continual enslavement of native peoples in Brazil, Brazil in the eighteenth century also saw a degree of economic change that reflected the importance of international competition to the slave world. The sugar plantations of north-east Brazil declined in importance from the 1710s, as sugar production from the West Indies instead became more important in supplying European markets. The British and French West Indies were closer to European markets than Brazil, which increased their competitiveness and profitability. This factor was accentuated by the degree to which the British and French domestic markets were more significant than that of Portugal. Moreover, the strength of British and French commerce emphasized their importance in reexport markets. This was an important aspect of the competition, and thus the search for comparative advantage, that the slave trade reflected. The sufferings of many served the cause of profit, with the dynamism of the latter driving new opportunities.

Slavery and International Finance

There was a major expansion of the Portuguese presence in the south of Brazil, where captaincies (administrative units) were founded for São Paulo (1709), Minas Gerais (1720), Goiás (1744), and Mato Grosso (1748). Gold and diamond extraction from the Brazilian province of Minas Gerais, where gold had been discovered in 1695, grew in significance, both rapidly and substantially. This growth produced at once a key demand for slaves in a different area from that of the sugar plantations, and a major new stimulus for the slave trade that, at the same time, helped to fund it. The crucial role of economic factors was amply displayed. Prior to the eighteenth century, black slaves were relatively infrequent in the region, but they were now imported in large numbers. Slaves worked both in mines and in tasks such as washing diamond-bearing rocks. Aside from in Minas Gerais, where the towns of Minas Novas and Diamantina were founded in 1727 and 1730 respectively, there were also gold deposits in Goiás and Mato Grosso. In the mining regions, the labor need for men, and the fact that these were new settlements, combined to lead to a situation in which men were often more than nine-tenths of the black population.[24]

Gold from Brazil was the major new source of bullion in the eighteenth-century global economy. Britain's close economic relationship with Portugal, a relationship that lasted into the twentieth century, ensured that these gold exports were highly significant for the liquidity of the British commercial system and economy. In part, this system and economy were therefore dependent on the slaves whose labors produced the gold, which indicated the multiple benefits of slavery to the British economy. Alongside the benefits stemming from commercial strength, political stability, and the continuity offered by a Parliament-funded national debt, Brazilian gold helped the British government to borrow at a low rate of interest. This was important to the funding of the British state—notably its war effort, which included the largest navy in the world.

Moreover, as a result of New World bullion resources, the West as a whole acquired an important comparative advantage in global trade. Asian powers might receive bullion for their products, such as tea and ceramics imported into the West from China, but access to bullion supplies ensured that Westerners were able to insert themselves into the non-Western world and to underwrite negative trade balances. The commercial position helped underscore the growth of Western financial and mercantile capital and organization.[25] This global economy focused on London. The British used gold not only to finance negative trade balances, particularly with China, India, Russia, and the Ottoman Empire, but also to subsidize allies and to hire foreign troops. Thus, the gold–slave economy was critical to British power.

Brazil and Angola

The Portuguese Atlantic was dynamic in a number of respects, and notably both economically and governmentally. The Marquês de Pombal, the chief minister from 1756 to 1777, strengthened the links between Portugal and its colonies while making the economies of the latter more flexible. This encouraged economic development, notably in Brazil, while his banning of the

Jesuits made the enslavement of indigenous Brazilians easier. In Brazil, from the late century, in another important geographical shift, sugar and coffee plantations near Rio de Janeiro became prominent. Rio replaced Salvador (Bahia), further north, as the capital of Brazil in 1763. Plantation goods exported from Brazil included sugar, tobacco, coffee, and, from the 1760s, cotton. These plantations were worked by slaves, and by 1800 there were over one million slaves in Brazil.

There was considerable differentiation in the slave trade in the South Atlantic as there was in the North Atlantic. The Portuguese bases in Angola supplied Minas Gerias and Rio de Janeiro with slaves, while West Africa supplied the sugar plantations of north-east Brazil that were also further north. This differentiation was an aspect of the wider range, but also specialization, of the Atlantic slave trade. It took about six weeks to sail from Elmina in West Africa to Salvador (Bahia), and about seven weeks to sail from Luanda in Angola to Rio de Janeiro, a key entrepôt in the slave trade.[26] The partnership between Europeans and elite Africans was crucial to this trade, with Luso (Portuguese)-African families, who spanned the Portuguese world of the Angolan and Guinea coasts and the African world of the interior, also having links into the plantation-owning families of north-east Brazil.[27] The use of words is again indicative: bases supplying plantations is a misleading innocuous phrase. Moreover, it does not focus on the degree to which labor needs led to a deliberate search for slaves.

Although the Brazilian plantations relied on African slave labor, and most slaves came from Angola, slave-raiding was still much practiced in Brazil at the expense of natives. Slave-traders from Sao Paulo, known as Paulistas or *mamelucos*, did a lot to lower native numbers. However, they were resisted by the *reducciones*, frontier settlements in Spanish America under Jesuit supervision where natives grew crops. This encouraged the Portuguese to close down the Jesuit settlements. Native slaves and forced labor were important in Amazonia for the collection of cacao, sarsaparilla (the dried roots of which are used as a tonic), and other forest products to the benefit of the colonial economy. However, in 1743–9, possibly half the native population of the Amazon valley fell victim to measles and smallpox.[28] This was another instance of the vulnerability of the native populations in the Americas to European illnesses.

Spanish America

Export growth from Spanish America was also linked to the intensification of slavery, which encouraged the import of slaves there. Moreover, in 1789, Spain allowed the free trade in slaves to some of its American colonies, including the Spanish West Indies and Venezuela. This was seen as a blow to both the regulated trade (*asiento*) and to illegal trade. Across the century, there was growth in the export of sugar and tobacco from Cuba, of cacao and sugar from Mexico, and of cacao, tobacco, cotton, coffee, sugar, and indigo from Venezuela. In contrast, there was no large-scale, slave-based plantation economy in Pacific Spanish America (modern Ecuador, Peru, northern Chile), nor in Argentina, where there was a Spanish colony focused on the estuary of the River Plate. The distances to European markets from both were too great to provide competitive advantage for such a plantation economy.

Qualifying arguments based solely on racialism, part of the black population of Latin America (as elsewhere) were not slaves. Moreover, there was a willingness to arm the free black people. They were increasingly recruited into the militia in Spanish America from 1764. Furthermore, the defense of the interior of the French colony of Cayenne (French Guiana) was entrusted to Native Americans and free black people, who were organized into a company of soldiers. In addition, in the 1760s and 1770s, black and mulatto (mixed black–white) Brazilians were recruited into companies of irregular infantry. This process was far less common in British America, in large part because the settlers, who were opposed to the arming of black people, were more powerful there, both in the West Indies and in North America.

The British Slave Trade

Brazil was a market for British slave-traders. There were also British and French sales to Spanish America, both legal and illegal, but for British, as for French, merchants, the core trade was that of selling slaves to their own colonies. The British were the most prominent in the slave trade. Between 1691 and 1779, British ships transported over 2 million slaves from African ports. London dominated the trade until the 1710s, when it was replaced by Bristol.[29] In turn, the developing port of Liverpool took the leading position from the 1740s. The regulatory framework that had maintained London's control had been dismantled in 1698, when the African trade was freed from the control of the Royal African Company. This legalized the position of private traders and made shifts in the relative position of ports far easier. The merchants of Bristol had long criticized the monopolistic role of London trading companies.

The value of liberalization, in the shape of a dismantling of regulations limiting trade, were to be seen at a different scale with the removal of restrictions on the Mozambique trade from South-East Africa in the second half of the century. The monopoly role of companies there was ended when the Portuguese colony of Mozambique in South-East Africa was removed from the jurisdiction of Portuguese India; and in 1786 the monopoly of the port of Mozambique was ended: other ports in the colony gained their own customs houses and began trading. This liberalization led to an expansion in the slave trade from Mozambique that was also related to the ability to expand the area in the interior from which slaves were obtained. This expansion rested on the purchase of slaves, and not on the territorial expansion of Portuguese Mozambique attempted earlier in the colony's history, and that was not resumed until the late nineteenth century.

In Britain, the shift from Bristol to Liverpool as the leading slave port was readily apparent. In 1725, Bristol ships carried about 17,000 slaves and, between 1727 and 1769, 39 slavers were built there. In contrast, by 1752, Liverpool had 88 slavers with a combined capacity of over 25,000 slaves. In 1750–79, there were about 1,909 slave trade sailings from Liverpool, 869 from London and 624 from Bristol.[30] Liverpool had better port facilities than Bristol, not least the sole wet dock outside London, and this dock was followed by a major expansion of dock facilities, with the opening of the Salthouse Dock (1753), St. George's Dock (1771), and Duke's Dock (1773). Liverpool became the key European site of the slave trade, a rapidly expanding city devoted to commerce. Money from the trade enabled investment in nearby transport links and manufacturing, and these further increased the importance of the port.

Most slaves were transported by the British to the West Indies, which were the center of the British slave economy. In an instructive variant on the overwhelming use of slaves for private owners, there were also "King's Negroes," slaves owned by the British state. The state purchased, hired, or obtained, by default (legal means notably bankruptcy) or as prize, slaves in large numbers for military service and in smaller numbers for naval service. In the army, slaves were under arms and also used as "pioneers"—i.e. workers for expeditionary forces and as labor to build, maintain, and service fortified positions.[31] The navy used slaves afloat, but principally ashore in order to man the naval yards that were crucial to the maintenance of British naval power, notably at English Harbour on Antigua, at Kingston and Port Antonio on Jamaica, and on Bermuda. At the same time, privately hired slave labor was more significant for these naval yards, let alone for the British West Indies as a whole.

Many slaves also went to British North America, and the slave trade to it was largely an eighteenth-century phenomenon. The number of slaves in the British colonies there rose from about 20,000 in 1700 to over 300,000 by 1763. This was particularly the case as, first, South Carolina and, then, Georgia were developed as plantation economies, supplementing those already developed on the Chesapeake. In South Carolina and, eventually, Georgia, rice became an additional plantation crop, reflecting the particular opportunities of the tidewater environment for cultivation. The bulk of the colonial economy that took part in overseas trade was located in the tidewater as opposed to further inland, and ships sailing directly to Europe moored in many of what now seem minor inlets. With their shallow draughts, the ships could readily service coastal and riverside plantations—for example, along the River York. To that extent, the continental mainland of North America was similar to the islands of the West Indies.

The slave trade was integral not only to the plantation regions, but also to the commercial economy and shipping world of the British Atlantic, crucial to entrepreneurial circles in Britain, and to the financial world there, and had an extensive range of influences elsewhere in Britain, particularly, but not only, in the ports. The large ports of Liverpool, Bristol, Glasgow, and London were directly and closely involved in the trade, either as the source of slave ships, or of the goods they carried, or of the resulting plantation goods. As in North America, many smaller ports were also involved. They included Barnstaple, Bideford, Dartmouth, Exeter, Lancaster, Plymouth, Poole, Portsmouth, Topsham, and Whitehaven. The last, now a relatively obscure town, sent about 69 slaving voyages to Africa between 1710 and 1769, most after 1750, and its merchants were probably the fifth most important group of slave traders in Britain in the latter period.[32]

The role of the smaller ports helped spread the impact of the slave trade on the British economy. However, many, including Barnstaple, Bideford, Dartmouth, Exeter, Plymouth, and Topsham, all ports in Devon, played only a minor role. This was a process eased by the extent to which many merchants and ships were involved in the trade on a temporary basis. Returns from slave-trading ventures were risky, but also sufficiently attractive to keep some existing investors in the trade and to entice new investors to join up. The returns could enable men of marginal status to prosper sufficiently to enter the merchant class.

Furthermore, the triangular pattern of Atlantic trade was practicable for many small-scale operators, as the outlay of funds required was less than for the trade to the more distant East Indies. This pattern involved, first, goods, both British manufactures and imports, such as East India Company textiles, shipped from Britain to Africa. Second, after they had been sold and the money or credit spent on buying slaves, slaves were taken from Africa to the New World. Third, with the profit from the sale of the slaves, colonial products, such as sugar and tobacco, were transported from the New World back to Britain.[33] The triangular trade depended on credit and the ability to wait for payment. Britain's financial strength provided both. The crucial role of finance, particularly of the sugar commission business, ensured that London, where the business centered, was as heavily involved in the slave trade as Liverpool from which there were more sailings.[34] Financial buoyancy was particularly important in long-distance trade in which new ships lasted only two or three voyages, financial returns were seriously delayed, and merchants, therefore, needed to obtain long-term credit on favorable terms.

The trade brought prosperity to British manufacturers. The export of goods to Africa as part of the triangular trade helped broaden the range of groups in British society who were interested in the slave trade and who, directly or indirectly, profited from its expansion. This increased the penetration of the slave trade in the British economy and in British society. Britain's largest industry, the woolen textile industry, benefited greatly from the development of the slave trade. The share of these textiles in the export of British-made goods to Africa rose from 6.5 percent in 1660 to 64.9 percent in 1693, a year in which European markets were greatly affected by war; and from then until 1728 never dropped below 47.1 percent. Moreover, the percentage of British exports to Africa produced in Britain rose from less than 30 percent in the 1650s and 1660s to about two-thirds by 1713–15.[35] This opportunity helped manufacturers to cope with periods of difficulty in their usual markets, and led to innovations, notably in the production of lighter cloth, as the African market was pursued.[36]

Africa was seen as a sphere of opportunity not least because trade to much of the rest of the world appeared blocked or limited by the role of other European states and their protectionism. A key instance was Spain's determination to restrict foreign trade with its colonies. British trade with much of the rest of the world seemed limited by the attitude and strength of non-European powers, or restricted by the entrenched position of monopolistic British chartered companies, such as the East India Company. In some parts of the world, the extent of opportunity was affected by the lack of any product worth obtaining in return for Western exports. The last, for example, limited Portuguese and Dutch interest in Australia. In his "Thoughts on the African Trade," published in 1730 in number 76 of the *Universal Spectator*, "J.L." attacked the position of the British East India Company in trade to the Indian Ocean and urged the establishment of a trade to the south-east coast of Africa, with textiles and pewter exported. This, he argued, would help British industry, whereas, he claimed, the trade to India did not bring such advantages. The suggestion did not bear fruit and South-East Africa did not become an area of economic significance for Britain until the nineteenth century: it was far more distant from Britain and the New World than West Africa and Portuguese interests in South-East Africa were well established. At the same time, J.L.'s article was indicative of the sense of Africa as a land of opportunity for British trade. This argument was later to be taken up by Abolitionists who claimed that the abolition of the slave trade would create major new opportunities for British exporters.

The British triangular trade offered considerable flexibility, so that, for example, when sugar became harder to obtain from the West Indies, Lancaster's traders found other imports in which to invest their proceeds from slave sales, particularly mahogany, rum, and dyewoods (wood yielding dyes), each of which was in demand in Britain. This flexibility enabled the traders to maximize their profits on each leg of their enterprise. Such a maximization was particularly important for marginal operators trading in goods and with markets where they faced domestic and international competition. Moreover, when competition did eventually make the slave trade less viable at Lancaster, the contracts and experiences forged by the African trade meant that other trading opportunities, both abroad and in Britain, were on offer to merchants.[37]

Profitability was not only an issue for British traders. The profitability of the colonial trade of La Rochelle, one of France's leading Atlantic ports, was affected by the wild fluctuations in slave-trade profits, as well as by wars and attendant colonial losses. As a result, merchant families there limited their business endeavors neither to maritime trade, nor to any one branch of it. Nevertheless, in France, long-distance trade was more profitable than private notarized credit, while being as profitable as, and also safer than, government bonds.[38] In addition, risk and solvency were serious problems in the slave trade from Angola to Brazil. Moreover, difficulties in attracting investment affected the Coriso Company established by Portugal in 1723 to export slaves to north-east Brazil. This company collapsed in 1725.

The British triangular trade across the Atlantic was not the sole commercial system that was developed to help finance and exploit slavery. Supplying food and other products to the slave plantations was also important. Serving the slave economies was part of the British commercial world, with the development of a trade in salt cod from Newfoundland, both to the West Indies and to Charleston, the port for South Carolina. Moreover, food was shipped from the Thirteen Colonies that were to become the basis of the USA to the British West Indies. The colonial contribution included slaving from Rhode Island: from the ports of Newport, Bristol, and Providence. Rhode Island lacked the agricultural opportunities offered elsewhere in the bigger colonies and produced rum that could be exported to Africa in return for slaves.[39]

Kendal, for example, was an English inland town that did not have close links to a major port; yet, in his *Tour of Scotland 1769* (1771), Thomas Pennant recorded how trade had encouraged industry there, with cloth manufactured from wool across northern England and Scotland, and then exported, via Glasgow, for the use of slaves in Virginia.

There were similar developments in other colonial systems. Some of the expansion in French exports to the West Indies was for the growing population of the French Indies. However, much of the production was reexported to the Spanish Empire, a much larger market. In 1741–2, the French city of Bordeaux exported over 8 million livres' worth of produce to the French West Indian colonies annually, principally wine and textiles. The comparable figure for 1753–5 had risen to over 10 million livres annually. The wealth of the slave economies ensured that the white settlers were well able to purchase goods exported from metropolitan France.

Individual merchants interacted and multiplied the opportunities for each, as well as the overall frequency, range, and impact of national trading systems. Moreover, each of the lands abutting the North Atlantic acquired a comparative economic advantage which could then be developed and exploited by entrepreneurs in a version of the division of labor applauded by the British economist Adam Smith in his pathbreaking *Inquiry into the Nature and Causes of the Wealth of Nations* (1776). Accumulated experience in the slave trade helped improve responsiveness to opportunities, lessened risk, and increased the efficiency of the trade. This efficiency included a responsive African supply system involving both Europeans buying slaves in West Africa and their African suppliers.[40] It was possible to expand supply to meet, or try to meet, changes in demand, and there was also a successful search for new sources of supply.

The slave trade, nevertheless, involved serious commercial risks, created, for example, by a lack of sufficient slaves or, alternatively, by the glutting of markets due to too many slaves arriving at once, and these issues greatly affected the price of slaves and thereby the profitability of voyages. The need to sail rapidly once slave cargoes had been obtained, so as to lessen risks from disease, meant that ship captains could reach the New World with a cargo of slaves that did not match the requirements of the purchasers. Moreover, the trade could be expensive to enter and the scale of the likely profits was unclear. Alongside evidence of considerable profits and a major contribution as a result by the slave trade as a whole to investment and economic growth in Britain, there are more modest estimates, and suggestions that, at the level of the individual voyage, the slave trade did not, on the whole, bring great profits, or sometimes even any. The serious risks, not least a shortage of slaves in West Africa (and therefore higher prices), disease during the Atlantic passage, and a temporary glut of slaves in the New World ports (and therefore low prices) were separate, but they created a cumulative sense of unpredictability. Concern about the profitability of the trade was a major factor in the pronounced variation in the number of voyages per year from individual ports. As an example of the benefits of scale that were so significant in capitalist systems, larger firms had a considerable advantage.[41]

Because the majority of British ships involved made only one voyage in the trade, there was no substantial, separate, specialized slave fleet. Partly as a result, many slave ships were not the large ships that tend to attract attention, but smaller vessels carrying a limited number of slaves and often with scant preparation. Most of the ships used fell within the 50–150 tonnage range.[42] Similarly, for the Portuguese trade, the large *galeras* were less important than the more modestly sized *bergantim*.

For Britain, as for other countries, the individual merchant and voyage is a key context for consideration,[43] although the UCL database of slave ownership and compensation suggests that slave ownership was more common than is usually assumed, and this extended to roles in shipping. The collapse of the role of the Royal African Company ensured that this was even more the case for the British trade. Most slaves were brought by private traders and, opposing the company, they pressed for the freedom to do so on their terms.[44] The company's bases in West Africa meant that it exercised a function of sovereignty in a sphere where the state did not wish to do so. However, the bases entailed a considerable financial and organizational burden. In 1730, the near-bankrupt company petitioned for a government subsidy for the maintenance of its forts and settlements. Indeed, from 1730 to 1747, in order to support this maintenance, the company received an annual subsidy from the government. However, even with this addition, the financial arrangement agreed in 1698 could not save the company's poor finances. In 1750, the company was abolished by the African Trade Act, which opened the trade to all British subjects willing to pay a fee. With the help of a block grant from Parliament, the company's bases were to be maintained by a successor company, the Company of Merchants Trading to Africa, which was controlled by the leading slave traders.[45] The poor financial state of the Dutch West India Company and the heavy cost of its West African operations, again of the support of bases, similarly led it to depend on Dutch government subsidies.

As an indication of the range of difficulties faced by slave traders, Bristol merchants in the early 1730s were hit by shortages of slaves, falling profits on colonial reexports as prices dropped, and deteriorating relations with Spain which created doubt about security and markets: war between Britain and Spain finally broke out in 1739. The number of slave voyages sailing from Bristol fell from 53 in 1738 to 8 in 1744; the war only ended in 1748 and British merchants were heavily exposed to privateering during it. In the case of Whitehaven, its merchants largely abandoned the slave trade after 1769. These points understandably tend to be overlooked in the modern discussion about the profits made from the slave trade and the related claims for compensation. However, alongside the constant fear of slave rebellion on the ships, the financial precariousness and general uncertainty of much of the trade may well have contributed to elements of its cruelty, both deliberate and unintentional. Indeed, this cruelty can be seen at the various levels of organization, notably both large-scale movement and smaller-scale, more improvised, activity.

The slave trade and slavery were also affected, both directly and indirectly, by the competition between rival, or potentially rival, British commercial interests over the opportunities created by the Atlantic world. In 1739, for example, British merchants supported agitation for the export of sugar from Britain's West Indian colonies direct to European markets, and not therefore via Britain, a move which was resisted by the British sugar refiners. This pressure against Britain's entrepôt role was designed to reduce costs and increase flexibility for merchants, but, as in this case, the pressure challenged British industrial interests involved in processing. These rivalries extended to include economic interests in the colonies, notably planters but also merchants and manufacturers.

Profitability was hit by the human cost of slavery, in the shape of the frequently (although not invariably) high death rate of slaves on the Atlantic crossing. With time, the percentage of slaves carried who died fell appreciably, but this was largely owing to shorter journey times across the Atlantic rather than to improved conditions, although a reduction in the crowding on ships was important.[46] Slaver captains were less concerned about slaves' survival than the slave owners were. In particular, as demonstrated in the case of the *Zong*, there was little interest in costly medical care, although this was also true for the crew. Keeping slaves imprisoned in crowded conditions in the fetid holds of the ships (as, earlier, in the coastal bases where the slaves were held), rather than permitting opportunities for exercise and sunlight on deck, was very bad for the health of the slaves and thereby hit the interest of the owners. However, such imprisonment responded to the concerns of captains worried about rebellions by the slaves. Many of the officers and crew involved in the trade also died, largely as a result of tropical diseases. These casualties, which made the slave trade unpopular among sailors, bring out anew the deadly nature of this business.

Developments in Africa

Meanwhile, the ships continued to sail and warfare in Africa among African powers continued to provide large numbers of slaves. The intensive nature of warfare in much of the Atlantic hinterland—for example, the long civil war in Kongo in what is now north-western Angola and western Congo, and the westward advance of the Lunda Empire in what is now eastern Angola, fed the slave trade, as did droughts and famines, which increased debt. Family members were sold or pawned into slavery to cancel debts. Such sales were also ways to raise income. The *sahel* belt to the south of the Sahara, moreover, saw widespread conflict. For example, in the valley of the River Niger, the Pashalik of Timbuktu declined under pressure from the Tuareg and was destroyed by them in 1787. This was an aspect of the longstanding rivalry between the urban-based cultures of the valley and the nomadic peoples of the Sahara Desert to the north. This rivalry, long linked to slaving raids by the latter, continues to the present day. Instability in the Niger Valley affected the trans-Saharan slave trade to North Africa and also provided slaves for the Atlantic trade. Further south, in the forest zone of West Africa, musketeers had largely replaced archers on the Gold Coast in the seventeenth century, and did so on the Slave Coast in the eighteenth. Asante in the former and Dahomey in the latter were both expansionist powers whose conflicts produced slaves.[47] Neither was frightened of Western slave-traders.

Alongside this supply-side account of why slaves were available, it is also pertinent to note the degree to which the slave trade helped transform African society, hitting earlier patterns of hierarchy and identity, and savaging agrarian communities. Trends in slave prices suggest that the external demand factor became more important from the mid-eighteenth century and, as a result, that developments and forces within Africa became relatively less important in the history of the regions affected.[48]

However, for most of the century, variations in the productivity of the trade, not least the time spent by ships off Africa, reflected African forces—namely, shifts in slave supplies within Africa, notably as a result of conflict there. Indeed, British terms of trade with West Africa fell heavily in the 1720s, and then again in the 1750s–70s, before rising in the 1780s and 1790s, although to a figure that was only about a third of that in the 1700s. From the last quarter of the century, Western advances in shipping and organization, notably larger ships, became important and helped lead to major gains in productivity in the slave and other trades.[49] These gains could take precedence over shifts in supplies within Africa, but proved most effective when they were combined with them.

In practice, both Africans and Europeans had agency, with Europeans intervening in the search for favorable links, while African warrior-rulers were also slave traders.[50] One of the readers of the first edition of this book wanted all reference to African agency removed. That would be completely mistaken.

Military developments provided a major element in the African context of the slave trade. In West Africa, there was a diffusion of Western arms without political control by the Western powers, although the traffic in firearms developed more slowly at a distance from the coast. Muskets, powder, and shot were imported into Africa in increasing quantities and were particularly important in the trade for slaves. For example, in the Senegal valley, gifts of guns were used to expand French influence. Moreover, Fort Bakel, in the east of modern Senegal, was founded by the French in 1697, and Fort Medine, in the west of modern Mali, in 1719. This expansion helped in the focusing of slave movements on France's coastal bases. Riverboat convoys on the River Senegal were highly important to the ability to use the river to link the interior to the slaving ports. Moreover, moving slaves by water was less arduous as well as quicker than marching them overland, and thus increased profitability. The Senegal valley led at the coast to the French slave base at Fort Louis (Gorée is further south), while that of the Gambia led to the British bases at Fort James and Albreda.

An emphasis on firearms puts the Westerners center-stage, at least in so far as supplies for them were concerned, but the role of firearms in African warfare should not be exaggerated. The Mahi, whose warfare was based on bows and arrows, successfully resisted Dahomian attack in the 1750s; Dahomey blocked the supply of guns to them. In addition, cavalry was more important than firearms to the north of the forest belt in West Africa. It was largely thanks to cavalry that the kingdom of Oyo (in modern north-east Nigeria) was able to defeat Dahomey and force it to pay tribute from the 1740s. The Lunda of eastern Angola, who spread their power in mid-century, particularly relied on swords. The successful forces in the *jihad*, launched by the cleric Usuman dan Fodio and the Fulani against the Hausa states in modern northern Nigeria in 1804, initially had no firearms and were essentially mobile infantry forces, principally archers, able to use their firepower to defeat the cavalry of the established powers. Their subsequent acquisition of cavalry, not firearms, was crucial in enabling them to develop tactics based on mobility, maneuver, and shock attack. By 1808, all the major Hausa states had fallen to the *jihadis*, indicating that firearms were not crucial for war winning in Africa.[51] This led to the establishment of the Sokoto state. Thus, the provision of firearms, shot, and gunpowder by Westerners was not necessarily essential for the local powers or to the slave trade. Much depended on the particular circumstances of local warfare, circumstances which owed much to the local environment. This *jihad* was more successful than any Western military activity in West Africa's interior until the second half of the nineteenth century.

Further emphasizing the extent to which Westerners did not dominate the situation, in West Africa they remained confined to coastal enclaves, and not always in a satisfactory fashion. British cannon drove off Dahomey forces that attacked their fort at Glehue in 1728. However, these forces had already captured the Portuguese fort there in 1727. In 1729 and 1743, the Dahomians succeeded in capturing the Portuguese fort at Whydah, where Brazilian tobacco was exchanged for slaves, and did so despite the availability of firearms in the fort. Western garrisons faced a number of problems, notably disease and neglect. Wooden gun-carriages rotted rapidly in the Tropics, just as did hemp and flax, while iron corroded and many men died of disease. The difficulties the Portuguese encountered in East Africa are instructive. They regained Mombasa in 1728, after a mutiny by African soldiers against Omani control. However, the Portuguese soon lost Mombasa again to the Omanis. It fell in 1729, with the garrison capitulating as a result of low morale and problems with food supplies. Thereafter, the Portuguese never regained the position. Further south, the Dutch, expanding east from Cape Colony, overran the Hottentots and Bushmen with few difficulties, but then encountered the more numerous Xhosa, leading to a war between 1779 and 1781. The Xhosa mounted serious resistance, but were defeated on the River Fish.

A major problem facing Westerners was a lack of knowledge about the African interior. This was only lessened from the close of the eighteenth century, with the Association for Promoting the Discovery of the Interior

Parts of Africa, or African Association, being founded in London in 1788. The Association sponsored exploration of the interior of Africa from the River Gambia. From 1802, the British government took over the sponsorship of the major expeditions. It supported Mungo Park who, however, died in 1806 negotiating rapids on the Niger. His *Travels in the Interior Districts of Africa* (1799) provided new information for mapmakers, as did William Browne's *Travels in Africa, Egypt, and Syria* (1800).

The British were not alone in exploration. In 1798, Francisco Lacerda e Almeida, the Portuguese Governor of Sena on the River Zambezi in Mozambique, who was concerned about the British conquest of Cape Colony from the Dutch in 1796, decided that the Portuguese needed to link their colonies of Mozambique and Angola, and thus span southern Africa. He reached Lake Mweru in central Africa, but died of disease there. Linking the colonies remained an unrealistic goal until the end of the nineteenth century when it was, anyway, frustrated in the 1880s and 1890s by British expansion northward from Cape Colony, part of the British effort to create a continual strip of control from the Cape to Cairo.

Trade with Africa was only possible thanks to the active cooperation of African rulers, which extended, on the part of King Agaja of Dahomey to an interest, in the 1720s, in permitting the establishment of Western plantations in Dahomey. These were intended as a supplement to, rather than a substitute for, the slave trade.[52] Cooperation was also the pattern for the Arab slavers based on ports such as Zanzibar, Kilwa, and Mombasa who played the key role in the Indian Ocean slave economy. The relationship between foreign slavers and Africans was often a complex one. Within the wider context set by the Western inability to coerce the Africans, there was an emphasis by Africans not simply on trade as a way to obtain slaves, but also on the provision, and therefore availability, of credit. This was more generally true of the Western-dominated Atlantic trading system. Credit was given to encourage coastal Africans to purchase slaves from the interior. This was a reminder that, alongside coercion and its possibilities, slaves at each stage were also obtained by commerce. Mutual trust was crucial to the credit system. There was also a reliance on the security provided by connections of the African traders who were left with the Western merchants, and themselves shipped as slaves if the traders did not deliver. The general assumption was that trust would be fulfilled. Indeed, although there could be violent episodes, mutually accepted rules came to prevail. Far from these rules being set by Western traders, they reflected the ideas of African elites as to what constituted appropriate enslavement and an acceptable trade in slaves. This care did not extend to the slaves who were traded. Captives from war, crime, or debt, they were treated as without family ties, right, or respect, and therefore as fit for sale. The misery of the slaves, however, did not mean that the Africans involved as trade lacked a key and potent role in the trade.[53]

At the same time, there was a general process of change across part of Africa under the impact of the Western presence. For example, longstanding patterns of trade in West Africa between the desert and the savanna that had predated the expansion of Islam across the Sahara from the ninth century were annexed to the Atlantic world. The export of slaves and gum arabic (a product used in textile manufacture that was the other major Western-controlled export from this region, and also used in the printing industry from the early 1800s) reconfigured local economies and interregional trade. In turn, the cloth and metallurgical industries in West Africa were hit by the export of Western goods provided in order to gain slaves. These Western exports were sold on terms likely to get the slaves.[54] The difficulties caused for West African manufacturers by imports from the West anticipated problems for the African economies that were to be more widespread in the nineteenth century.

Resistance

There was a degree of African resistance to slavery, entailing action against both African slave hunters and Westerners. The latter extended to 493 known risings on slave ships, risings which were especially common in the late eighteenth century.[55] The prospect of risings on the ships helped account for the confining harshness of slave conditions and ensured that their crew was considerably larger—usually twice—than on merchant ships of a similar size. This number of sailors affected the profitability and logistics of the trade. From this perspective, African resistance, or its threat, actively molded the trade and limited the numbers transported. To that extent, the parallel with Jewish risings against the Holocaust is not too close. In each case, resistance has been stressed of late, and this emphasis is appropriate in ensuring that slaves and Jews are not seen simply as passive objects. However, the Jewish risings had less of an impact on the Holocaust because the Germans were so much stronger.

The Western position was far more powerful as far as the slaves in the Americas were concerned than it had been in Africa. Nevertheless, in the Americas there were still both slave risings and the more widespread and frequent problems created by escaping slaves. In 1739, in the Stono rising in South Carolina, 100 slaves rose and killed 20 colonists, before being defeated by the militia and their Native American allies. The determination of some of the slaves to escape to Spanish Florida helped prompt the rising. The Spanish authorities in St. Augustine had announced that any slave who fled to Spanish Florida would receive freedom in return for militia service and conversion to Catholicism, both of which represented a rejection of British power. The Spaniards also built Fort Mose, a community for free African-Americans. After the suppression of the rising, a British advance in 1740 led to the destruction of Fort Mose. In turn, the Spaniards advanced north on Georgia in 1742 with a force that included free black people as well as Spanish regulars and Native Americans. The Spaniards sought to win support from the Native Americans and black people, encouraging the latter to raid the British plantations. The Spanish advance was, however, blocked by British resistance.

Although the events of 1739–42 did not lead to the breakdown of the British slave economy in the region, they indicated the extent to which rivalry between the Western powers could well provide opportunities for slave opposition. This was to be clearly demonstrated in the War of American Independence (1775–83) and then, more clearly, because the fate of the slave rising on the French colony of Saint-Domingue from 1791 coincided, from 1793, with that of war between Britain and France. As a reminder of the more general significance of points arising from the consideration of the slave trade and slavery, this situation matched that which helped lead to success for revolutions in the Americas (and later elsewhere) against imperial rule.

Although their scale varied, there were slave risings in the British West Indies: on Antigua in 1735–6, Jamaica in 1742, 1760–1, 1765, and 1776, Montserrat in 1768, and Tobago in 1770–4, and this is not an exhaustive list.[56] It was similar for the colonies of other powers. Risings were usually brutally suppressed, and followed by savage punishments and harsh retribution in order to deter fresh upheavals. A Frenchman in Louisiana in the early eighteenth century recorded the burning to death of three black slaves who had been conspiring with Native Americans to overthrow the colony.[57] Alongside risings, there were also murders of individual overseers, as well as many suicides on the part of slaves.

Action by the white authorities and settlers lessened the chance of risings, and there were none in colonial Virginia and Maryland. This action included control over firearms, efforts to prevent slaves plotting, and declarations of martial law. Major efforts were made to ensure that slaves were unable to coordinate action, except in very small areas. Legislation was a product of concern about slave risings. The Stono rising of 1739 led the South Carolina Assembly to increase the duty on new slave imports, in order, at that point of crisis, to limit their number. Slaves born to other slaves in North America were regarded as far less of a threat. Hierarchy played a key role in controlling slave resistance. This was not only the hierarchy of white control but also the role of highly privileged slaves, slave drivers, and slave overseers, as well as of free black planters and slave owners.

Flight was a more common form of resistance. It was seen across the slave world but with particular areas of intensity. In Jamaica, the British mounted unsuccessful military expeditions against the Maroons—runaway slaves who controlled much of the mountainous interior. The prospect of slaves running away to join the Maroons led the Admiralty to argue that a naval base on the north shore of the island was a poor idea. The failure of the British expeditions, against elusive opponents in the difficult terrain of Jamaica, was followed, in 1738 and 1739, by treaties that granted them land and autonomy, creating a form of coexistence between the British authorities and the Maroons.

This was a key aspect of the extent to which the New World led to a great variety of circumstances for Africans, one in which their capacity to mold their life was crucial. Particularly in Jamaica and Brazil, but also elsewhere, there were organized communities of fugitive slaves, including many in Cuba as well as the "Bush Negroes" in Surinam. In Saint-Domingue, Maroons were made more threatening by the power attributed to belief in Vodou, a means of drawing on the spiritual power of the dead in order to wield power over slaves, offsetting that of their white masters. International borders and jurisdictions added a further dimension, making cooperation to lessen slave flight less likely. The *Briton*, a London newspaper, in its issue of December 11, 1762, argued that gaining St. Augustine in Florida from Spain would prevent "the desertion of our Negro slaves" from Georgia. That, indeed, happened in the subsequent peace treaty.

Slaves who did not flee could, if circumstances were propitious, also engage in social protest and labor bargaining. In Barbados during the eighteenth century, the insurrectionist attitudes that had led to plans for revolts in 1649, 1675, and 1692 were replaced on the part of slaves by an emphasis on limited protest that was designed to secure the amelioration of circumstances.[58] This contrasted with the other British islands such as Jamaica where circumstances were harsher. In general, the situation was less favorable than on Barbados.

There was also the possibility of slave risings as part of broader-based opposition, as in 1730 when there was white support for a rising on the coast of Venezuela against the attempt by the Caracas Company to end smuggling. The rising was suppressed by troops. Slave agitation in the region was to become an aspect of a wider current of uncertainty. From the 1740s, stories circulated in the region that the Spanish government had issued an order to free all slaves throughout the empire, but was being thwarted by the local authorities. An enslaved healer called Cocofío spread such disruptive and inaccurate reports from the 1770s to his death in 1792. In the 1790s, such reports were to be fanned by reports of slaves being freed in the French West Indies. Spanish concern about the situation in the 1790s led to the end by the Spanish government to the decrees granting freedom to runaway slaves from rival powers; these slaves were now seen as a threat, rather than as a source of manpower[59]—for example, as slaves fleeing South Carolina and Georgia had long been regarded in Spanish-ruled Florida. The idea of runaway slaves as a threat represented a major change to the previous emphasis on accumulating and controlling or, if appropriate, rewarding them.

White concern about slaves reflected multiple anxieties, and anxieties at the individual as well as the collective level. It was easy to misinterpret poor slave health or understandable complaints from black people about conditions as active defiance, and to react harshly accordingly. This assumption greatly contributed to the coercive nature of the slave economy and the reliance on force, a situation made worse by the arbitrary nature of authority. Beatings were very common. Visiting Charleston with his brother John in 1736, Charles Wesley noted in his journal: "It was endless to recount all the shocking instances of diabolical cruelty which these men (as they call themselves) daily practice upon their fellow-creatures; and that on the most trivial of occasions." He commented on a slave woman beaten unconscious and revived, before being beaten again, and having hot candle wax poured on her skin. Her offense was overfilling a teacup.[60] Aside from often routine and widespread physical violence, there was considerable verbal abuse as well as the contempt of gesture.

It is notable that there was a long gap between the Wesley brothers' confrontation with slavery in South Carolina and John Wesley's highly influential anti-slavery pamphlet *Thoughts Upon Slavery* that appeared in 1774, by which time the influence on Wesley of Granville Sharp and Anthony Benezet had dramatized the transatlantic origins of Anglo-American Abolitionism. Benezet (1713–84), a Quaker, emigrated to America in 1731, becoming a Pennsylvania teacher. He established an evening school in Philadelphia for slaves and published extensively on the evils of slavery, including *A Caution and Warning to Great Britain and her Colonies on the Calamitous State of the Enslaved Negroes* (1767) and *Some Account of Guinea, with an Enquiry into the Slave Trade* (1771), works that influenced Thomas Clarkson, a prominent Abolitionist. Granville Sharp (1735–1813), a polymath, became conspicuous in Britain for supporting slave causes from the mid-1760s, publishing *A Representation of the Injustice ... of tolerating Slavery* (1769).

Slave Life

The circumstances of slave life varied greatly.[61] For example, some slaves, at least, became skilled men in the plantations, and hence acquired a self-respect that owners could recognize as a mutual advantage. Slaves frequently acted as the overseers of other slaves, although they never acquired the authority and power given to particular slaves in the Ottoman Empire. There were also efforts to ameliorate conditions. Alongside the brutal situation in Surinam reported in *Candide* by Voltaire and cited earlier in the chapter, came the *Rule on the Treatment of Servants and Slaves* introduced there by the Dutch in 1772 in order to counter slave unrest. This *Rule* was to be enforced by the Fiscal, an official to whom slaves had the right of appeal. In general, however, government regulation was limited, infrequent, and remote. Instead, owners and traders were the key players.

More profoundly, alongside the individual level, regional variations in plantation economies had important consequences for the slave trade and for the circumstances of slave lives. As far as British America was concerned, owing to the varied demands of tobacco and rice cultivation, and to related economic and social characteristics, slaves in the Chesapeake were more affected by white life, living in close proximity to owners in relatively small farms. In the Chesapeake, the slave population approached demographic self-sufficiency (could reproduce itself), as early as the 1720s. In contrast, in the rice lands of South Carolina, there were fewer, but larger, plantations, and the percentage of slaves was greater. Moreover, the higher death rate in South Carolina compared to the Chesapeake ensured that, although by the 1750s the slave workforce could reproduce itself, there were more imported African slaves compared with the American-born slaves who were more significant in the Chesapeake. As a consequence, in South Carolina, allowing for the difficulties in assessing slave attitudes, slaves were more autonomous and more influenced by African culture and material life than in the Chesapeake, and relations between slaves and whites remained more antipathetical than where they lived in closer proximity.[62]

The slave situation in Jamaica, where slaves were also treated harshly, was more similar to South Carolina than the Chesapeake. Among both slaves and white people in Jamaica, the majority were immigrants, with the enslaved crucial to the sugar and rum production that was responsible for much of the population. A free black and mulatto population existed, but was outnumbered by slaves.[63]

The diaries of Thomas Thistlewood, who was a slave owner and agricultural manager in west Jamaica from 1750 to 1786, indicate clearly that he treated his slaves cruelly, not least abusing his female slaves sexually, a practice depicted in the 1798 caricature *Tit Bits in the West Indies*.[64] Sexual exploitation was a product of white male power as well as the extent to which far more African women crossed the Atlantic than European ones prior to the nineteenth century.

Similarly, on Sir William Stapleton's profitable sugar estate on the island of Nevis in the West Indies, there were very exploitative working arrangements, and in the 1730s the black population there fell by nearly 25 percent. In the British Windward Islands—Dominica, Grenada, St. Vincent, and Tobago—in the 1760s and early 1770s, there was harsh treatment that did not encourage family life and reproduction, in large part because it was easier to buy new slaves than to raise children to working age. The resulting regime, with far too little food provided to slaves, led to high slave death rates and to low slave fertility. The severity of treatment contributed to flight and suicides, while the causes of low female fertility worried slaveowners and moralists. In practice, much was due to the difficulties of working sugar, but owners and many commentators preferred to opine on sexual mores, blaming those who were in fact poorly treated.[65]

There were variations in the treatment of slaves within the British West Indies. For example, master–slave relations appear to have been less brutal on the British Leeward Islands than in Jamaica.[66] The slave situation in Brazil was more similar to that in South Carolina, Jamaica, and the Windwards than that in the Chesapeake or the Leewards. The slave populations of Brazil, Cuba, Jamaica, Trinidad, and British Guiana had negative growth rates and were not approaching demographic self-sufficiency, and this continued to be the case in the 1820s.

Sexual exploitation, shocking in itself and a matter of even greater concern in light of current mores, was an established part of male settler life and led to criticism by moral commentators. Overlapping with exploitation that could be callous, cruel, and involve rape, there were informal unions, interracial consensual sex, and illegitimate children. The gender dimension was very pronounced, as the sexual behavior of white women was treated in a far less liberal fashion than that of white men, a situation that matched the lack of political rights of the women and their limited control over property, which was the situation in the colonies as in the European homelands. Race and gender were not the sole issues. There was also a matter of social control by the wealthy over the poor, the owner over the worker, such that Barbados black planters after the age of slavery had extramarital relations with female plantation laborers.[67]

The situation was dynamic. The growing need for slaves in South Carolina and Georgia, where settlement expanded considerably, helped ensure that the balance of the British slave world shifted, so that, by 1775, there were more slaves in British North America than in the British West Indies. In part, this situation also reflected the amelioration of the slave position in the Chesapeake, which led there to a rising rate of natural increase, which was to become more apparent in the nineteenth century. Rising prices for slaves in British North America from mid-century indicates that demand was outstripping supply. Differing needs for slaves were a key aspect of the variety of the slave economies and fed through into slave flows. The Ibo shipped from the Bight of Biafra were favored by Virginia planters whose focus was on tobacco cultivation, but the Ibo were seen as insufficiently strong by Carolina and Georgia planters who preferred Bambara and Malinke from Senegambia as well as Angolans. There was similar variety offshore. In the British colony of Bermuda, tobacco cultivation had been replaced by the cultivation of food in order to supply other colonies as well as shipping. Bermuda also became an important center for trade. Comprising close to half of the working population, slaves there were crucial not to a plantation economy, but to a broad-based seafaring and agriculture one. They provided labor, including skilled labor, as in shipbuilding, and were also a key aspect of society. Nearly nine-tenths of white households owned slaves.[68] The employment of slaves in Bermuda, and their related conditions, were very different from the situation in the Bahamas, let alone Barbados.

Racism

The treatment of slaves was related, often very closely, to the racism of the period. The contrast between white workers sent to the Americas as indentured laborers, not slaves, and Africans moved there in floating prisons, as slaves, not indentured labor, reflected more than racism, but racism was essentially linked to the contrast. Racist attitudes were far from restricted only to the unlearned, but, as today, could be found across society. Religious explanations of apparent differences between races, whose genesis was traced back to the sons of Adam, with black people as the children of the cursed Ham, were important. Although there were prominent exceptions—for example, the Prefects of the Capuchin Missions in Congo (important Catholic missionaries) in the mid-seventeenth century—most churches found no difficulty in reconciling slavery with the attempt to extend the Christian message to slaves. This can readily be seen with the Dominican and Jesuit missionaries in the French Caribbean or with the Society for the Propagation of the Gospel in Foreign Parts (SPG), an influential British organization.

The views and practices of this Anglican body are important because it was a key part of the establishment of religious activity and culture in the eighteenth-century British Atlantic. Moreover, the SPG was highly influential to the development of white attitudes in North America and the West Indies. Founded in 1701, with Thomas Tenison, Archbishop of Canterbury from 1695 to 1715, the senior cleric in the Church of England, as its first President, the SPG focused on Britain's American and Caribbean colonies and founded a college in Barbados in 1716. While arguing that slaves should not be treated brutally, the SPG presented slavery as compatible with the social order and argued that the Christian instruction of slaves was crucial to sustaining the stability of this order. This view was maintained at the highest level of the Church of England. Edmund Gibson, Bishop of London from 1723 to 1748, and, from 1716 until 1736, the key ecclesiastical adviser to the first minister, Sir Robert Walpole, issued a pastoral letter in 1727 in which he supported both slave conversion to Christianity and the biblical sanctioning of slavery. Gibson's use of freedom is instructive:

The freedom which Christianity gives is a freedom from the bondage of Sin and Satan, and from the dominion of men's lusts and passions and inordinate desires; but as to their outward condition, whatever that was before, whether bond or free, their being baptized, and becoming Christians, makes no manner of change in it ... And so far is Christianity from discharging men from the duties of the station and conditions in

which it found them, that it lays them under stronger obligations to per-
form those duties with the greatest diligence and fidelity.[69]

The attitude of the SPG was harshly displayed on Barbados where, in 1710,
it was bequeathed the valuable Codrington plantation. The subsequent
treatment of the slaves on this plantation was scarcely different from the
norm. Public whippings were designed to maintain order, runaways were
branded with the word "Society," children were forced to work, and families
were separated by sales. Closely identified with slave-holding, the SPG found
it difficult to win slave converts. In mid-century, as an alternative, it tried to
use Western-educated African schoolmasters in missionary schools. This
also proved unsuccessful.[70]

Alongside discussion about divine intention, biological explanations for
slavery played a role. Influential writers argued in favor of polygenism—the
creation of different types of humans, with this difference being the case
from the outset when, allegedly, the types were created differently. This thesis
led to suggestions that black people were not only a different species, but
also related to great apes, such as orangutans. These suggestions were linked
to the argument that, although black people were allegedly inherently inferior, they were particularly adapted to living in the tropics.

Physical attributes, particularly skin color, attracted much attention, with
Montesquieu and Buffon explaining color as due to exposure to the tropical
sun. The ability of black people to cope better than white people with diseases in the tropics was believed to exemplify an inherent difference that was
linked to a closeness to animals that lived there. This was held, by white
people, to justify slavery. The argument that bile was responsible for the
color of human skin, advanced as a scientific fact by writers in Antiquity,
was repeated without experimental support by eminent eighteenth-century
scientists, including Buffon, Feijoo, Holbach, and La Mettrie. An Italian
scientist, Bernardo Albinus, proved to his own satisfaction in 1737 that a
black person's bile was black, and in 1741 a French doctor, Pierre Barrère,
published, in his *Dissertation sur la cause physique de la couleur des Nègres*,
experiments demonstrating both this and that the bile alone caused the
black pigment in their skin. This inaccurate thesis, which was discussed in
The Gentleman's Magazine, the leading British journal, the following year,
won widespread acclaim, in part thanks to an extensive review in the *Journal
des Savants* in 1742. The theory played a major role in the prevalent mid-
century belief that black people were another species of man without the
ordinary organs, tissues, heart, and soul. In 1765, the chief doctor in the
leading hospital in Rouen, Claude Nicolas Le Cat, demonstrated that
Barrère's theory was wrong. However, Le Cat was generally ignored and
Barrère's arguments continued to be cited favorably.[71] They accorded with a
hierarchical classification of humanity that served the interests of the slave
trade as well as the views of Western society.

The alleged nature of African society was also deployed as an argument in favor of slavery. For example, in his *A New Account of Some Parts of Guinea and the Slave Trade* (1734),[72] William Snelgrave, an active English slave trader from 1704, defended the trade on the grounds that being sacrificed was the alternative for prisoners. This was an argument that was to be adopted later in the century by opponents of the abolition of the slave trade.

It is instructive in this context that the most successful sentimental critique of slavery published in early eighteenth-century Britain related not to a black slave but to a Native American. Native Americans were not generally treated with the racism shown to black people. The tale of "Inkle and Yarico" published in the leading London periodical, the *Spectator*, on March 13, 1711, was one of humanity affronted and morality breached by slavery:

Mr. Thomas Inkle, an ambitious young English trader cast ashore in the Americas, is saved from violent death at the hands of savages by the endearments of Yarico, a beautiful Indian maiden. Their romantic intimacy in the forest moves Inkle to pledge that, were his life to be preserved, he would return with her to England, supposedly as his wife. The lovers' tender liaison progresses over several months until she succeeded in signaling a passing English ship. They were rescued by the crew, and with vows to each other intact, they embark for Barbados. Yet when they reach the island Inkle's former mercantile instincts are callously revived, for he sells her into slavery, at once raising the price he demands when he learns that Yarico is carrying his child.

It is significant that Barbados, a leading center of the slave trade, is chosen as the site for this change. The popularity of the story of Inkle and Yarico, the subject of a frequently performed play by George Colman the Younger, first produced in 1787, reflected the way in which the transoceanic world could provide a setting for moral challenges.[73] Slavery was to become the most important instance of this process. Already, prior to the late eighteenth century, there was an ambivalence in the treatment of black people, notably of those in domestic European settings, where there were few of them, particularly in northern Europe. The portrayal of black people could be sentimental, and the harsh world of their work and lifestyle was generally ignored, although this criticism could also be advanced of the extensive artistic treatment of the European peasantry. Joseph Wright's intimate and gentle portrayal of *Two Girls with their Black Servant* probably depicts the daughters of a merchant in Liverpool where Wright worked in 1769–71. The service shown here seems agreeable and bore little reference to the far grimmer nature of reality, not least the role of Liverpool as the leading slave port.

In the first half of the eighteenth century, any reluctance to introduce slavery and the slave trade arose essentially from prudential calculations.

For example, slavery was banned by the Trustees of the new colony of Georgia in 1735, not so much because of hostility to slavery as due to the wish to base the colony on small-scale agrarian activity rather than aristocratic plantations. Defensive considerations played a role and the law was entitled "An Act for rendering the Colony of Georgia more defencible by prohibiting the importation and use of Black Slaves or Negroes into the same." The newly established colonists, however, opposed the measure from the outset, not least because they had to watch their South Carolinian neighbors getting rich using cheap slave labor, while they could barely eke out a living. The Malcontents, the leading political faction in Georgia that opposed the Trustees, published several pamphlets calling for legalizing the slave trade, but they were unsuccessful. The Stono Rebellion in South Carolina in 1739 was cited by the Trustees as evidence of the danger of slavery.

The Trustees, however, finally caved in in 1750, arguing that, since Spain had been unsuccessful in 1742 when it invaded Georgia from Florida during the War of Jenkins' Ear (1739–48), it was safe to import slaves. This compromise followed those on other restrictions regarding issues such as land tenure and alcohol consumption. It was an aspect of the collapse of the Trustees' position and they surrendered their charter in 1752.[74] Their objective of a colony of virtuous small farmers had not been realized, but it prefigured the goal of Thomas Jefferson, America's president from 1801 to 1809, and a Virginian who, however, combined an aspiration for such a society with slavery. The Georgia Trustees had wanted silk or wine as their monoculture, but neither worked out very well. The economy of Georgia only came close to being a success after the Trustee era, and real wealth in Georgia had to wait for King Cotton.

By the end of the eighteenth century, most Western advanced opinion no longer regarded black people as a different species of man, but as a distinct variety. This interpretation, monogenesis—the descent of all races from a single original group—was advanced by Johann Friedrich Blumenbach (1752–1840), a teacher of medicine at the University of Göttingen. In 1776, he published *De Generis Humani Varietate*, an influential work of racial classification that went through several editions. However, the misleading assessment of the inherent characteristics of non-Westerners, combined with the association of reason with Western culture, still encouraged a hierarchy dominated by the Westerners, and thus a treatment of others as inferior, a treatment to which the association of blackness and slavery greatly contributed. Thus, the reconceptualization of racism was in part related to the slave trade.

Although monogenesis can be seen as a benign theory that could contribute to a concept of the inherent brotherhood of man that was voiced during the Enlightenment and especially in the period of the American and French Revolutions, it was also inherently discriminatory. Blumenbach assumed the original ancestral group to be white and argued that climate, diet, disease, and mode of life were responsible for the developments that led to the creation of different races. Considering the relative beauty of human skulls, so as to determine the history of the human species, Blumenbach claimed that that of the Caucasian girl was most beautiful and the original production of nature. Aesthetics, therefore, was deployed alongside racial science in support of notions of Western superiority. Moreover, race came to be understood as a biological subdivision of the human species rather than, as originally, a people or single nation linked by a common origin.[75]

Separately, but as part of the equation, to supporters of slavery, an acceptance of black people as fully human did not preclude slavery, as they were presented as degraded by their social and environmental backgrounds.

Racial characteristics and developments were understood in terms of the suppositions of Western culture, and this led to, and supported, the hierarchization already referred to. This situation also tended to be true of the developing idea of cultural relativism. However, subversive themes could be offered, as by the leading British painter William Hogarth in his *Analysis of Beauty* (1753). He pointed out that "the Negro who finds great beauty in the black females of his own country, may find as much deformity in the European beauty as we see in theirs."[76]

Among "advanced" thinkers, notions of brotherhood, however, were subordinated to a sense that Enlightenment and Revolutionary ideas and movements originated within the Western world. Irrespective of the nobility of individuals, their societies appeared deficient and defective, and thus inferior. This was seen in writing on history and sociology—for example, William Robertson's highly influential *History of America* (1777), in which the natives of the New World were presented as less advanced than the conquering Europeans. Slaves were thus considered in a broader context of a ranking of societies in which the West was superior and prevailed.

The American Revolution

The combination of racist views with the economic interest of the white colonists helped ensure that, however significant for the future, the cause of freedom in the case of American liberty did not extend to the slaves. Indeed, dedicated to the leading French radical, Jean-Jacques Rousseau, the third edition of *The Dying Negro* (1773) by the British radical Thomas Day criticized the American Patriots for supporting slavery. In addition, the Hessian soldiers sent to fight the Americans on behalf of George III felt that the American treatment of their slaves formed a hypocritical contrast with their claims of the equality of man. The use of black soldiers in conflict against white people proved a particularly sensitive issue and notably in America during the American Revolutionary War. In practice, the British did not use black soldiers on any scale, and certainly not on that to be seen in the West Indies in the French Revolutionary and Napoleonic Wars. About 9,000 African Americans served in units of the Continental Army and about the same number fought for George III.[77]

However, in the 1770s, both American independence and the use of black troops underlined the dynamic nature of circumstances and the extent to which the slave trade and slavery were affected by external factors. More generally, prior to abolition, there was no "steady state" of slavery, but, rather, from the outset, a continual procession of adaptation to changing circumstances. Slaves were treated as commodities in this adaptation, which reflected the degree to which they were at once a labor force deployed, alongside capital and technology, to ensure such adaptation, but also a labor force that was owned and coerced to this end.

Many of the black people who fought for the British, and of those who fled to the British lines from slavery, far from rejoining the new American country, crossed the Atlantic at the end of the war in 1783. Some went to London. Others were dispatched to Nova Scotia, to strengthen the British presence. Promises of free land there, however, were often not kept, while the land itself that was available was poor. From both London and Nova Scotia, many of the black people went to the new colony of Sierra Leone in West Africa, where they found the promise of free land was matched by the reality of disease. These were slave journeys in reverse: from slavery in the former Thirteen Colonies to a new diaspora of ex-slaves. Deaths from disease and a troubled time after landing was the fate of many, even though they were not traveling as slaves.[78] There were problems, moreover, elsewhere. On the Bahamas, black Loyalists faced attempts at reenslavement by white Loyalists seeking to introduce an economy based on cotton plantations, a pattern seen in America. As governor from 1787, John, 4th Earl of Dunmore, resisted this process.[79] Earlier, as the last royal governor of Virginia, he had armed slaves in 1775–6.

With racism and its ideology and practice of exclusion still strong,[80] the American constitution did not end slavery. Indeed, the 1787 Constitutional Convention prohibited the government from interfering with the slave trade for 20 years and this was one of the Clauses that was unamendable. This provision was partly intended to enable the slave states to compensate for the lack of imports during the War of Independence and for the extent of black flight during the war. Moreover, South Carolina and Georgia both expected a future need for slave imports. The majority of the black population (which was 19.3 percent of the national population in 1790) was in the South. In contrast, slavery was slowly abolished in the states of the North, although, with the exception of Massachusetts, this process did not involve the emancipation of slaves, but, rather, as in New York under legislation of 1799,[81] the freeing of slave children born thereafter once they reached maturity. This, however, was a slow process.

Conclusion

Alongside the continuation of slavery, there was also much harsh treatment of other workers, in the West and elsewhere, that led to comparisons with slavery. Traveling through the Alps from Füssen in Bavaria to Innsbruck in Austria in 1787, Adam Walker wrote of the women he saw: "I sincerely pity them, they are such slaves as I have heard the Negroes in the West Indies described. No uncommon sight to see them threshing corn, driving wagons, hoeing turnips, mending the highways."[82] It would not have benefited the slaves to know that such comparisons could be made, nor that they were part of a more dynamic economic system in which consumerism, capital accumulation, and investment in industrialization were all linked.[83] For example, profits accumulated in Glasgow from sugar and tobacco trading helped fund the development of the chemical industry in west-central Scotland and also increased the liquidity of Scottish banks.

This was an aspect of the extent to which the British Atlantic stood out from the other European Atlantics in terms of the combined degree and intensity of the processes of exchange and linkage.[84] The slave trade was a major part of these important processes, while the profits of the slave economy helped in the development of British industrialization. However, the Industrial Revolution owed more to coal, which was produced by paid British workers and not by slaves, than it did to sugar.

At the same time, from the 1770s and, notably, 1780s, alongside a greater reluctance by slaveowners in Britain to demonstrate their position, not least by advertising about runaway slaves, there was increased criticism of slavery among evangelical Christians and progressive intellectuals. For the latter, anti-slavery was both a mission to fulfil the divine plan, and a seeming and desirable compromise between radical tendencies and acceptance of the social order. Evangelical moral universalism was fused with the secular universalism of the Enlightenment and natural rights. The *Encyclopédie* (1751–65), the manual of French enlightened thought, had been characterized by contradictory or tentative views, alongside criticism—for example, by the Chevalier de Jaucourt in his entry on slavery.[85]

In contrast, Abbé Raynal's influential *Histoire philosophique et politique des établissements et du commerce des Européens dans les deux Indes* (1770) became, especially in its 1774 and 1780 editions to which the French intellectual Denis Diderot made important contributions, a channel for the expression of progressive ideas, such as anti-slavery. The *Histoire* was widely read by Western reformers and was translated into English. Within a quarter-century of the appearance of the *Histoire*, the slave trade was to be under severe pressure from advanced opinion in the Atlantic world, while one of the leading slave colonies, Saint-Domingue, was consumed by revolution.

Notes

1 Voltaire, *Candide* (1759), chapter 19.
2 T. Smollett, *The Adventures of Roderick Random* (London, 1748), chapters 24, 36, 45, 46.
3 T. Smollett, *The Adventures of Peregrine Pickle* (London, 1751), chapter 52.
4 T. Smollett, *Travels Through France and Italy* (London, 1766), chapter 4.
5 Ibid., chapters 32, 14.
6 S.P. Newman, "Freedom-Seeking Slaves in England and Scotland, 1700–1780," *English Historical Review*, 134 (2019), 1167–8.
7 D. Eltis and D. Richardson, eds., *Extending the Frontiers: Essays on the New Transatlantic Slave Trade Database* (New Haven, CT, 2008).
8 P. Gauci, ed., *Regulating the British Economy, 1660–1850* (Farnham, UK, 2011).
9 J.M. Postma, "A Reassessment of the Dutch Atlantic Slave Trade," in *Riches from Atlantic Commerce; Dutch Transatlantic Trade and Shipping, 1585–1817*, ed. Postma and V. Enthoven (Leiden, 2003), 137.
10 R. Ross, *Cape of Torment: Slavery and Resistance in South Africa* (London, 1983).
11 P.O. Hernaes, *Slaves, Danes, and African Coast Society: The Danish Slave Trade from West Africa and Afro-Danish Relations on the Eighteenth-Century Gold Coast* (Trondheim, 1998).
12 R.L. Stein, "Measuring the French Slave Trade, 1713–1792/3," *Journal of African History*, 19 (1978), 520–1.
13 R.L. Stein, "The French Sugar Business in the Eighteenth Century: A Quantitative Study," *Journal of Business History*, 22 (1980), 14.
14 L.M. Rupert, *Creolization and Contraband: Curaçao in the Early Modern Atlantic World* (Athens, GA, 2012), 204.
15 D.J. Hamilton, *Scotland, the Caribbean and the Atlantic world 1750–1820* (Manchester, 2005), 181.
16 P. Woodfine, *Britannia's Glories: The Walpole Ministry and the 1739 War with Spain* (Woodbridge, 1998).
17 Anon., *A Letter from a Merchant of the City of London to the Right Honorable William Pitt Esq* (2nd ed., London, 1757), 44–5.
18 Lagau, French Consul in Hamburg, to Ministry of the Marine, May 13, 1748, AN. AM B7 365.
19 Lagau to Ministry of the Marine, January 5, 1750, AN. AM. B7 375.
20 Lagau to Ministry of the Marine, January 26, 1750, AN. AM. B7 375.
21 BL. Add. 36797 fol. 1.
22 M.P. Dziennik, "'Till these Experiments be Made': Senegambia and British Imperial Policy in the Eighteenth Century," *English Historical Review*, 130 (2015), 1132–6.
23 A.J. O'Shaughnessy, *An Empire Divided: The American Revolution and the British Caribbean* (Philadelphia, PA, 2000), 166.
24 L.W. Bergad, *Slavery and the Demographic and Economic History of Minas Gerias, 1720–1888* (Cambridge, 1999).
25 L. Neal, *The Rise of Financial Capitalism: International Capital Markets in the Age of Reason* (Cambridge, 1990).
26 S.B. Schwartz, *Sugar Plantations in the Formation of Brazilian Society: Bahia, 1550–1835* (Cambridge, 1985), 182.
27 J.C. Miller, *Way of Death: Merchant Capitalism and the Angolan Slave Trade, 1730–1830* (Madison, WI, 1988); for a critical evaluation, M. Candido, "Capitalism and

The Slave Trade at its Height 115

Africa: Revisiting *Way of Death* Thirty-Five Years after its Publication," *American Historical Review*, 127 (2002), 1439–48; P. Mark, *"Portuguese" Style and Luso-African Identity: Precolonial Senegambia, Sixteenth–Nineteenth Centuries* (Bloomington, IN, 2002); Mariana P. Candido, *An African Slaving Port and the Atlantic World: Benguela and Its Hinterland* (Cambridge, 2013).

28 D. Sweet, "Native Resistance in Eighteenth-Century Amazonia: The 'Abominable Muras' in War and Peace," *Radical History Review*, 53 (1992), 58.

29 D. Richardson, ed., *Bristol, Africa and the Eighteenth-Century Slave Trade to America: II, The Years of Ascendancy, 1730–1745* (Gloucester, 1987).

30 D. Richardson, "The British Empire and the Atlantic Slave Trade, 1660–1807," in *The Oxford History of the British Empire, II: The Eighteenth Century*, ed. P.J. Marshall (Oxford, 1998), 446; J.E. Inikori, *Africans and the Industrial Revolution in England: A Study in International Trade and Economic Development* (Cambridge, 2002), esp. 479–82.

31 R.N. Buckley, *The British Army in the West Indies: Society and the Military in the Revolutionary Age* (Gainesville, FL, 1998); B.S. Dyde, *The Empty Sleeve: The Story of the West India Regiments of the British Army* (London, 1997).

32 D. Richardson and M.M. Schofield, "Whitehaven and the Eighteenth-Century British Slave Trade," *Transactions of the Cumberland and Westmorland Antiquarian and Archaeological Society*, 102 (1992), 183–204, esp. 195.

33 For the French equivalent, R. Lemesle, *Le Commerce colonial triangulaire: XVIIIe-XIXe siècles* (Paris, 1998).

34 R.B. Sheridan, "The Commercial and Financial Organization of the British Slave Trade, 1750–1807," *EcHR*, 2nd ser., 11 (1958–9), 249–63, esp. 263; J.A. Rawley, *London: Metropolis of the Slave Trade* (Columbia, MO, 2003).

35 J.E. Inikori, *Africans and the Industrial Revolution in England: A Study in International Trade and Economic Development* (Cambridge, 2002), 407–8, 416, 513, 518–19.

36 M.D. Mitchell, "Three English Cloth Towns and the Royal African Company," *Journal of the Historical Society*, 13 (2013), 447.

37 N. Tatterfield, *The Forgotten Trade, Comprising the Log of the Daniel and Henry of 1700 and Accounts of the Slave Trade from the Minor Ports of England, 1698–1725* (London, 1991); M. Elder, *The Slave Trade and the Economic Development of Eighteenth-Century Lancaster* (Preston, 1992).

38 G. Daudin, "Profitability of Slave and Long-Distance Trading in Context: The Case of Eighteenth-Century France," *Journal of Economic History*, 64 (2004), 144–71.

39 J. Coughtry, *The Notorious Triangle: Rhode Island and the African Slave Trade, 1700–1807* (Philadelphia, PA, 1981); A. Jones, "The Rhode Island Slave Trade: A Trading Advantage in Africa," *Slavery and Abolition*, 2 (1981), 225–44.

40 D. Eltis and S.L. Engerman, "Fluctuations in Age and Sex Ratios in the Transatlantic Slave Trade, 1663–1864," *EcHR*, 2nd ser., 46 (1993), 321.

41 J.E. Inikori, "Market Structure and the Profits of the British African Trade in the Late Eighteenth Century," *Journal of Economic History*, 41 (1981), 745–76, esp. 774–5. For the argument that there was a reasonable rate of return, W. Darity, Jr., "Profitability of the British Trade in Slaves Once Again," *Explorations in Economic History*, 26 (1989), 380–4. A clear summary is offered by K. Morgan, *Slavery, Atlantic Trade and the British Economy, 1660–1800* (Cambridge, 2000), 36–48.

42 W. Minchinton, "Characteristics of British Slaving Vessels, 1698–1775," *Journal of Interdisciplinary History*, 20 (1989), 53–81, esp. 74.

43 J.C. Appleby, "'A Business of Much Difficulty': A London Slaving Venture, 1651–1654," *The Mariner's Mirror*, 71, 1 (1995), 3–14; R. Harris, *The Diligent: A Voyage*

through the Worlds of the Slave Trade (New York, 2002), re. a French ship that in 1731–2 sailed from Vannes on France's Atlantic coast, to Africa and then to Martinique before returning to Vannes; R. Damon, *Joseph Crassous de Médeuil: 1741–1793, marchand, officier de la Marine royale et négrier* (Paris, 2004).

44 Anon., *A True State of the Present Difference between the Royal African Company and the Separate Traders; showing ... the Advantages and Reasonableness of an Open Trade to Africa; and lastly the Danger of an Exclusive Trade, not only to the Traders of South and North Britain, but to our American Plantations* (London, 1710).

45 W. Pettigrew, *Freedom's Debt: The Royal African Company and the Politics of the Atlantic Slave Trade, 1672–1752* (Chapel Hill, NC, 2014).

46 H.S. Klein and S.L. Engerman, "Slave Mortality on British Ships, 1791–1797," in *Liverpool, the African Slave Trade and Abolition*, eds. R. Anstey and P.E.H. Hair (Liverpool, 1976), 113–25.

47 R.L. Roberts, *Warriors, Merchants, and Slaves: The State and the Economy in the Middle Niger Valley, 1700–1914* (Stanford, CA, 1987); S.P. Reyna, *Wars Without End: The Political Economy of a Precolonial African State* (Hanover, NH, 1990); J.K. Fynn, *Asante and its Neighbours, 1700–1807* (London, 1971).

48 J. Miller, "Worlds Apart: Africans' Encounters and Africa's Encounters with the Atlantic in Angola, before 1800," *Actas do Seminário Encontro de Povos e Culturas em Angola* (1995), 274; Richardson, "Prices of Slaves in West and West-Central Africa: Toward an Annual Series, 1698–1807," *Bulletin of Economic Research*, 43 (1991), 21–56, esp. 47.

49 D. Eltis and D. Richardson, "Productivity in the Transatlantic Slave Trade," *Explorations in Economic History*, 32 (1995), 465–84, esp. 480; H.A. Gemery, J.S. Hogendorn, and M. Johnson, "Evidence on English-African Terms of Trade in the Eighteenth Century," *Explorations in Economic History*, 27 (1990), 157–78, esp. 170.

50 S. Strickrodt, *Afro-European Trade in the Atlantic World: The Western Slave Coast c.1550–c.1885* (Woodbridge, 2015).

51 J.P. Smaldone, *Warfare in the Sokoto Caliphate* (New York, 1977); R. Law, *The Oyo Empire c.1600–c.1836: A West African Imperialism in the Era of the Atlantic Slave Trade* (Oxford, 1977).

52 Law, "King Agaja of Dahomey, the Slave Trade, and the Question of West African Plantations: The Mission of Bulfinch Lambe and Adomo Tomo to England, 1726–32," *Journal of Imperial and Commonwealth History*, 19 (1991), 138–63.

53 P.E. Lovejoy and D. Richardson, "The Business of Slaving: Pawnship in Western Africa, c. 1600–1810," *Journal of African History*, 42 (2001), 67–89.

54 J.F. Searing, *West African Slavery and Atlantic Commerce: The Senegal River Valley, 1700–1860* (Cambridge, 1993); D.R. Wright, *The World and a Very Small Place in Africa* (Armonk, NY, 1997).

55 S.A. Diouf, ed., *Fighting the Slave Trade: West African Strategies* (Athens, OH, 2003).

56 V. Brown, *Tacky's Revolt: The Story of an Atlantic Slave War* (Cambridge, MA, 2020).

57 G.M. Sayre, ed., *The Memoir of Lieutenant Dumont* (Chapel Hill, NC, 2012), 250.

58 H. Beckles and K. Watson, "Social Protest and Labour Bargaining: The Changing Nature of Slaves' Responses to Plantation Life in Eighteenth-Century Barbados," *Slavery and Abolition*, 8 (1987), 272–93, esp. 275.

59 L.M. Rupert, *Creolization and Contraband: Curaçao in the Early Modern Atlantic World* (Athens, GA, 2012), 203–5.

60 I.A. Brendlinger, *Social Justice through the Eyes of Wesley: John Wesley's Theological Challenge to Slavery* (Guelph, Ontario, 2006), 14–15; M.J. Fuentes, *Dispossessed Lives: Enslaved Women, Violence, and the Archive* (Philadelphia, PA, 2016).

61 R. Soulodre-La France, "Socially Not So Dead! Slave Identities in Bourbon Nueva Granada," *Colonial Latin American Review*, 10:1 (2001), 87–103.

62 P.D. Morgan, *Slave Counterpoint: Black Culture in the Eighteenth-Century Chesapeake and Lowcountry* (Chapel Hill, NC, 1998).

63 J.P. Greene, *Settler Jamaica in the 1750s: A Social Portrait* (Charlottesville, VA, 2016).

64 T. Odumosu, *Black Jokes, White Humour: Africans in English Caricature, 1769-1819* (London, 2017).

65 K. Paugh, *the Politics of Reproduction: Race, Medicine, and Fertility in the Age of Abolition* (Oxford, 2017); S. Turner, *Contested Bodies: Pregnancy, Childrearing, and Slavery in Jamaica* (Philadelphia, PA, 2017).

66 T. Burnard, *Mastery, Tyranny, and Desire: Thomas Thistlewood and His Slaves in the Anglo-Jamaican World* (Chapel Hill, NC, 2004); R.B. Sheridan, "The Condition of the Slaves in the Settlement and Economic Development of the British Windward Islands, 1763–1775," *Journal of Caribbean History*, 24 (1990), 121–45, esp. 141–2; K. Mason, "The World an Absentee Planter and his Slaves Made: Sir William Stapleton and his Nevis Sugar Estate, 1722–1740," *Bulletin of the John Rylands University Library of Manchester*, 75 (1993), 103–31, esp. 31; N.A. Zacek, *Settler Society in the English Leeward Islands, 1670–1776* (New York, 2010). See also Klein, *Slavery in the Americas: A Comparative Study of Virginia and Cuba* (Chicago, IL, 1967).

67 B. Bush, "White 'Ladies', Coloured 'Favourites' and Black 'Wenches'; Some Considerations on Sex, Race and Class Factors in White Creole Society in the British Caribbean," *Slavery and Abolition*, 2 (1981), 245–62.

68 M.J. Jarvis, *In the Eye of All Trade: Bermuda, Bermudians, and the Maritime Atlantic World, 1680–1783* (Chapel Hill, NC, 2010).

69 E. Gibson, *Two Letters* (London, 1727), 10–11.

70 T. Glasson, *Mastering Christianity: Missionary Anglicanism and Slavery in the Atlantic World* (Oxford, 2012).

71 G.S. Rousseau, "Le Cat and the Physiology of Negroes," *Studies in Eighteenth-Century Culture* (1973), 369–86.

72 R. Law, "The Original Manuscript Version of William Snelgrave's *New Account of Some Parts of Guinea*," *History of Africa*, 17 (1990), 367–72.

73 F. Felsenstein, ed., *English Trader, Indian Maid: Representing Gender, Race, and Slavery in the New World: An Inkle and Yarico Reader* (Baltimore, MD, 1999).

74 J.R. Hertzler, "Slavery in the Yearly Sermons Before the Georgia Trustees," *Georgia Historical Quarterly*, 59 (1975), 118–26.

75 N. Hudson, "The 'Hottentot Venus', Sexuality, and the Changing Aesthetics of Race, 1650–1850," *Mosaic*, 41:1 (March 2008), 31; N. Hudson, "From 'Nation' to 'Race': The Origin of Racial Classification in Eighteenth-Century Thought," *Eighteenth-Century Studies*, 29 (1996), 258; R. Wheeler, *The Complexion of Race: Categories of Difference in Eighteenth-Century British Culture* (Philadelphia, PA, 2000).

76 Cited in D. Dabydeen, "References to Blacks in William Hogarth's *Analysis of Beauty*," *British Journal for Eighteenth-Century Studies*, 5 (1982), 93.

77 W. Hotton, *Liberty is Sweet: The Hidden History of the American Revolution* (New York, 2021).

78 S. Schama, *Rough Crossings: Britain, the Slaves and the American Revolution* (New York, 2006); A.X. Byrd, *Captives and Voyagers: Black Migrants across the Eighteenth-Century British Atlantic World* (Baton Rouge, LA, 2008).

79 C.W. Troxler, "Uses of the Bahamas by Southern Loyalist Exiles," in *The Loyal Atlantic: Remaking the British Atlantic in the Revolutionary Era*, eds. J. Bannister and L. Riordan (Toronto, 2012), 185–207.

80 R.G. Parkinson, *The Common Cause: Creating Race and Nation in the American Revolution* (Chapel Hill, NC, 2016).

81 A. Zilversmit, *The First Emancipation: The Abolition of Slavery in the North* (Chicago, 1967).

82 A. Walker, *Ideas Suggested on the Spot in a Late Excursion* (London, 1790), 108.

83 R.A. Austen and W.D. Smith, "Private Tooth Decay as Public Economic Virtue: The Slave-Sugar Triangle, Consumerism, and European Industrialization," *Social Science History*, 14 (1990), 95–115.

84 D. Armitage and M.J. Braddick, eds., *The British Atlantic World, 1500–1800* (Basingstoke, 2002), 247; E. Mancke and C. Shammas, eds., *The Creation of the British Atlantic World* (Baltimore, MD, 2005).

85 B.M. Saunderson, "The *Encyclopédie* and Colonial Slavery," *British Journal for Eighteenth-Century Studies*, 7 (1984), 15–37, esp. 37.

5 Abolitionism

Le Rodeur, a French slave ship, took on 160 slaves in April 1819 at Bonny to the east of the Niger delta, but, in the confined hold, infectious disease hit hard, first with eye problems and then dysentery. Crowded in the noxious hold among their urine and excrement, the slaves found their water ration cut from eight ounces a day to half a wine glass. Allowed on deck for exercise, some threw themselves into the sea, which led to the others being continually confined. Contagion spread to the crew and by the time the boat had reached Guadeloupe many of the slaves and crew had lost their sight in one or both eyes, and 31 of the slaves had been thrown into the Atlantic:

> When the sun is down, if our row is not finished we get flogged. I received thirty lashes, as did Joe. We are taken to the stocks at night, and flogged next morning. We told the manager the work was too much, that we had no time to get our victuals, and begged him to lessen the task: this was the reason we were flogged.

The nineteenth century witnessed the end of legal slave trading in the Atlantic world. This was not a process free from considerable opposition nor from serious difficulties. However, the Abolitionist cause succeeded, first against the slave trade and then against slavery, and this marked a major change in the Atlantic world. The debate over Abolitionism, and therefore the arguments for and against, can be traced back into the seventeenth century in the case of the Quakers, but became far stronger in the later eighteenth. Among the key Abolitionist currents were religious pressure and secular idealism, each of which were international movements.[1] The first was particularly important in the Protestant world, although there had always also been a significant current of Catholic uneasiness about, and sometimes hostility to, slavery. Abolitionism should be located in both religious and secular contexts. The two were generally closely linked for, with the exception of revolutionary France in the 1790s, there was only limited expression of atheistical views. Indeed, with the breach with Christianity, there was in

DOI: 10.4324/9781003457923-5

part an effort to offer not atheism but, instead, an alternative form of religion. More commonly, religious and secular progress were thought of as part of the same process, and notably so in Britain, the leader, first in the slave trade and then in its abolition.

Denmark

In Protestant Europe, the slave trade was abolished first by Denmark in 1792. This was achieved by government decree (without an Abolitionist campaign), although the law did not come into force until 1803. In the meanwhile, the slave trade to the Danish West Indies (now the American Virgin Islands) was opened to all foreign nations, and their population was built up from about 25,000 slaves to about 35,000. In part, it was believed in Denmark that Britain and France would soon abolish the trade and would then seek to prevent other powers from participating, which proved to be an erroneous expectation in the short term but not in the long term. The Slave Trade Commission in Denmark also argued that slave conditions on the islands would improve if imports were banned, as it would be necessary to look after the slaves in order to encourage them to reproduce. In short, slavery would become less reprehensible, a view also taken by British Abolitionists.[2] In addition, elsewhere in the Protestant world, criticism of the slave trade and slavery developed in the Netherlands,[3] as in Britain. Protestants were more prone than Catholics to criticize slavery.

Britain

Changes in Britain were more important, because of her imperial position as the leading European imperial and naval power, due to her key role in the slave trade and in the Caribbean slave economy, and thanks to her potential influence on other states. The sources of opposition to the slave trade and to slavery were various in terms of those involved and their varied Abolitionist sympathies and action. Quakers in Barbados and Philadelphia had begun debating the morality of slave owning from the late seventeenth century, with the Quakers of Germantown in Pennsylvania producing a written protest in 1688 that rejected the idea that race justified enslavement. These arguments came to affect Quaker views in Britain and then views across a wider range. The London Yearly Meeting of Quakers in 1761 disavowed slave trading. Christian assumptions about the inherent unity of mankind and concerning the need to gather Africans to Christ played a major role in influencing British opinion. As a reflection and source of commitment, missionary activity among slaves became significant from the mid-eighteenth century. The Methodist leader John Wesley, who had seen much cruelty when he visited South Carolina, notably Charleston, in 1736,[4] strongly attacked both slavery and the slave trade in his *Thoughts upon Slavery* (1774) although he was friends with slave owners and accepted donations from them.[5] In 1791, Wesley was to send William Wilberforce a dying message urging him to maintain the Abolitionist cause. As a key leader of religious renewal, Wesley was a highly influential figure, and not only for Methodists. In polite British society, by the early 1770s, both slavery and the slave trade were increasingly seen as morally unacceptable, and in a culture in which social rank and manners were increasingly assessed in terms of a politeness and sensibility in which a widely diffused morality played a key role.[6]

Commercial benefits from the abolition of the slave trade were also predicted by some commentators. Malachy Postlethwayt argued, in his *The Universal Dictionary of Trade and Commerce* (1774), that the trade stirred up conflict among African rulers and thus obstructed both British trade and "the civilising of these people,"[7] the latter very much understood in Western terms. Belief in the benefits of slavery became far stronger in the early nineteenth century, not least as British merchants looked for new export markets, particularly in the face, first, of the Napoleonic attempt to weaken Britain by banning her trade with Europe and, subsequently, of continental protectionism directed against British exports—for example, the German *Zollverein* (Customs Union) established in 1834. Furthermore, the massive expansion of British industry meant that there were more goods for sale, as well as a desire for raw materials. Thus, a very different basis for trade with Africa was proposed.

Alongside religious and economic strands to Abolitionism, there were also judicial ones. Lord Mansfield's ruling, in the case of James Somerset in 1771–2, that West Indian slave owners could not forcibly take their slaves from England made slavery unenforceable there and was matched by a reluctance among Londoners to help return runaway slaves. The legal status of slavery in Britain, however, was still unclear. In his opinion, delivered on June 22, 1772, in a case that had become (and remains) a cause célèbre, Mansfield made the distinction that, while slavery was not illegal in England, the court would not recognize a slave owner's domain over a slave. The decision made the important point that *habeus corpus* also protected slaves. While Mansfield had framed the precise question to be decided as narrowly as he could, and in a way that attempted to reconcile his humane principles with his reluctance to create new law by interfering with the slave trade, others interpreted his judgment in a broad fashion. Mansfield in effect asked whether colonial slavery laws could be enforced in England and made it clear that his opinion did not involve the legality of the slave trade. He argued that "the state of slavery is of such a nature, that it is incapable of being introduced on any reasons, moral or political," but only as a result of deliberate legislation, which was indeed the case with colonial slavery laws. Despite the judgment, the recapture and deportation of escaped slaves continued in

Britain, as did the sale of assets that included slaves. Nevertheless, by the early 1790s, Mansfield's narrow interpretation of the law had given way to a broader one, and the deportation of escaped slaves had ended.[8]

Drawing on a variety of strands, Abolitionist sentiment, directed against the slave trade, became more overt from the 1780s, and was an aspect of the reform pressure and religious revivalism of the period. Indeed, the move in Britain to end the slave trade can appear very abrupt. A society that in 1770 was fully prepared to accept, and profit from, being the leading slave trader in the world, was seriously prepared, within a quarter-century, to curtail, if not end it, and did so just over two decades later. This is indeed surprising, although other abrupt shifts in belief, and in political and social attitudes, have occurred in world history—for example, the Protestant Reformation.

In the case of Britain, the disruptive impact of the American Revolution was probably more significant in affecting attitudes than contemporaries allowed for. This impact operated on a number of levels, but principally in making transformative change seem a possibility, as with the revolution itself. In Britain, moreover, there was a call for reform after the crisis of failure in the War of American Independence (1775–83) and in response to political instability in the early 1780s. There was concern about the nature, viability, and fate of the British Empire as a whole and a strong sense of national decline. The popular reception of Edward Gibbon's *Decline and Fall of the Roman Empire* (1776–88) captured this mood, as did the trial of Warren Hastings, Governor-General of British India, for corruption, and also George III's support for moral rearmament in Britain in the shape of backing for admonitions about public conduct. Gibbon's work invited attention to the theme of imperial transience and decay. Drawing on the example of Imperial Rome, wealth, especially from India, was seen in Britain in the 1780s as a source of political corruption and of pernicious, effeminate luxury. Whereas empire was to be regarded as a site and source of manliness in the late nineteenth century, it was a source of anxiety in the 1780s. This anxiety increasingly affected the response to the West Indies and the slave trade. In particular, like the new British Empire in India, as well as that of Imperial Rome (and in contrast to the North American Empire that had been lost where the majority of settlers were of British origin), the British Empire in the West Indies had no ethnic underpinning and was clearly imperial. Such attitudes encouraged unease and pressure for change.

While Quakers and others played a prominent role in Abolitionism, they were effective in part because of a more widespread shift in popular opinion in Britain. Abolition became, and remained, a popular campaign in Britain. In 1787, the Society for Effecting the Abolition of the Slave Trade, a national lobbying group, was established. Children also boycotted sugar.[9] Under William Pitt the Younger, Prime Minister from 1783 to 1801, and from 1804 until his death in 1806, the government, in the peace years of 1783–93, was particularly concerned with fiscal reform and in the negotiation of trade treaties with foreign states. This reform policy focused on a drive for national regeneration after defeat in the War of American Independence.

Moves to end or regulate the slave trade readily accorded with this drive. National lobbying, combined with government attitudes, helped lead to the Dolben Act of 1788, by which conditions on British slave ships were regulated. However, the ready responsiveness of the slave trade to the change in circumstances—its adaptiveness—was speedily demonstrated. The restriction on the number of slaves that British ships could carry led to an increase in the costs to the slave trader and encouraged the use of larger ships. In 1789, a less rigorous, but similar, act was passed by the Dutch, and in 1799 British restrictions were strengthened.

Abolitionist sentiment rapidly gathered pace in the late 1780s, and became a broadly grounded movement, one that reached out across British society. Abolitionism affected the arts, leading to the production of visual and literary images of the horrors of slavery, such as the medallion of the Society for the Abolition of the Slave Trade designed by William Hackwood and manufactured at Josiah Wedgwood's factory, the source of fashionable pottery. This was a potent fusion of politics and art, with the art in question providing a clear message. There was also a mass of pamphlet literature and discussion of the issue in humanitarian novels. Information on the slave trade and the colonies was sought by Abolitionists, leading to Thomas Clarkson's *The Substance of Evidence of Sundry Persons in the Slave Trade* (1788). The information accumulated by Clarkson and others is still of great use. The prolific Abolitionist literature struck both Evangelical Christian and rational notes. Key works included Clarkson's *An Essay on the Slavery and Commerce of the Human Species, Particularly the African* (1786), the translation of a Latin prize dissertation at Cambridge, and Thomas Burgess's *Considerations on the Abolition of Slavery and the Slave Trade, Upon Grounds of Natural, Religious, and Political Duty* (1789). Much is correctly made of the Methodist and Evangelical campaigns and arguments for Abolition, but, in addition, Burgess, the leading Anglican (in America, Episcopalian) advocate of Abolition, wrote one of the most powerful and comprehensive attacks on slavery. At the time of writing, he was a Fellow of Corpus Christi College, Oxford, and a Prebendary of Salisbury Cathedral. A key figure in the English religious establishment, Burgess was later Bishop of St David's and then Salisbury. He advanced religious arguments for ending slavery, but also encompassed wider philosophical and political arguments. Burgess's book commented on the "good effects, which would follow from the abolition of slavery, in preparing the way for the diffusion of Christianity and civilization in Africa, a liberal and extended communication between Africa and Europe, and the discovery of the interior parts of Africa."

Another moral argument was that the consumption of luxuries was responsible for the iniquities of the slave trade and, in turn, benefited from it. This was the theme of *An Address to the People of Great Britain, on the Utility of Refraining from the Use of West Indian Sugar and Rum* (1791) by William Fox. This argument led to the abstention movement of 1791–2, one in which women were prominent. Indeed, Abolitionism was the first major political campaign in Britain in which women played a highly prominent role.[10] Sugar was boycotted and alternative cooking recipes were published.

In opposition to this pressure, the Society of West India Merchants and Planters sponsored pamphlets in defense of the slave trade. There were pro-slavery arguments that reflected clear racism.

The controversy reached into the distant corners of the country. On April 7, 1789, for example, the *Leeds Intelligencer* reported the collection of £18 for supporting the application to Parliament for repeal of the trade "raised by voluntary contributions in a small part of the high end of Wensleydale ... The contributors (being chiefly farmers) were informed of the injustice and inhumanity of the slave trade by pamphlets circulated previous to the collection." Across the country, the press resounded with the battle—for example, *Swinney's Birmingham and Stafford Chronicle* of May 1791 included a poem by "H.F." praising William Pitt the Younger's recent parliamentary speech on the slave trade.

Pressure to abolish the trade was hindered by the importance of the West Indies to the British economy, as well as by the opposition of King George III and the House of Lords to Abolitionism. Both were keen supporters of property rights and established practices, while George's son William, later William IV (r. 1830–7) was caricatured by James Gillray in *Wouski* (1788) for taking a black lover in the Caribbean.[11]

In 1788, Dolben's Bill to regulate conditions on British slave ships was bitterly opposed in the Lords by the Lord Chancellor, Edward, Lord Thurlow, a favorite of the king and the most prominent conservative minister, while other ministers such as Charles, Lord Hawkesbury, the President of the Board of Trade and another royal favorite, and Thomas, Viscount Sydney, the Home Secretary, offered more muted opposition. On April 19, 1791, William Wilberforce's motion to bring in a parliamentary bill for the abolition of the trade was defeated in the House of Commons by 163 to 88. In moving the motion, Wilberforce had declared:

It is a blessed cause, and success, ere long, will crown our exertions. Already we have gained one victory; we have obtained, for these poor creatures, the recognition of their human nature, which, for a while, was most shamefully denied. This is the first fruit of our efforts; let us persevere and our triumph will be complete. Never, never will we desist till we have wiped away this scandal from the Christian name, released ourselves from the load of guilt, under which we at present labour, and extinguished every trace of this bloody traffic, of which our posterity, looking back to the history of these enlightened times, will scarce believe that it has been suffered to exist so long a disgrace and dishonor to this country.[12]

The following year, many Abolitionist petitions were presented to Parliament, as they had also been in 1788. In 1792, a motion for Abolition in 1796 was passed in the Commons by 151 to 132. Moving the motion in the Commons on April 2, 1792, Wilberforce had declared:

My conviction of the indispensable necessity of immediately stopping this trade remains however as strong and unshaken as ever … I cannot but believe, that the hour is at length come when we shall put a final period to the existence of this unchristian traffick … I never will desist from this blessed work.[13]

In the final speech in the Commons' debate, Pitt declared:

This great point is gained; that we may now consider this trade as having received its condemnation; that its sentence is sealed; that this curse of mankind is seen by the House in its true light; that that, the greatest stigma on our national character which ever yet existed, is about to be removed.[14]

However, the Lords, where conservative views were stronger, postponed the matter by resolving to hear evidence, which was a delaying tactic. In 1793, Wilberforce's motion to hasten the actions of the Lords was rejected, as was that to abolish the supply of slaves to foreign powers. In 1794, the latter motion passed the Commons, but was rejected in the Lords. In 1795, the Commons refused leave to bring in a bill for abolition of the slave trade, and in 1796 the bill was rejected on a third reading (all bills required three readings to pass the Commons). In 1797, 1798, and 1799, efforts failed. In 1804, the measure passed the Commons, but was defeated in the Lords, while in 1805 it failed the second reading in the Commons.

In part, repeated opposition to Abolition more specifically reflected the conservative response in Britain to reform agitation after the French Revolution, which broke out in 1789, and notably so after Britain went to war with France. Alongside popular pressure to end the slave trade, there was a populist tone to opposition to Abolition. In William Dent's cartoon "Abolition of the Slave Trade, or the Man the Master," published in London on May 26, 1789, colonial produce is shown waiting for a purchaser because its price has gone up, while a slave in Western clothes beats a semiclothed white man, saying "Now, Massa, me lick a you, and make you worky while me be Gentleman—curse a heart." In a world turned upside down, white people are depicted at work in the sugar fields, while black people feast under the words "Retaliation for having been held in captivity." These warnings from a menacing imaginary future were similar to those advanced in 1753 by the eventual successful opponents of the Jewish Naturalization Act. As a reminder of international competition, a foreigner in Dent's cartoon remarks, "By gar den ve sal have all de market to ourselves, and by underselling we sal send Johnny Bull's capitall and revenue to le Diable," while a Briton comments, "Why, if I have my rum and sugar and my tobacco at the old price—I don't care if the slave trade is abolished."[15]

Part of the opposition to Abolitionism derived from the continued conviction that slavery was compatible with Christianity, and also that slavery was sanctioned by its existence in the Old Testament of the Bible. For example, the Dutch Reformed Church, the established church in Cape Colony, the colony centered on Cape Town, argued that slaves were not entitled to enter the Church, and that conversion to Christianity would not make slaves akin to Europeans because they had been born to slavery as part of a divine plan. The Dutch settlers opposed missionary activity among their slaves; the same was true of British and other plantation owners in the West Indies. By 1800, there were nearly 17,000 slaves in Cape Colony. They were brought by sea from the Indian Ocean, largely from the Portuguese colony of Mozambique.

France

The French Revolution, which broke out in 1789, was linked to a secular idealism that embraced Abolitionism as one of its themes. In February 1788, *La Société des Amis des Noirs* had been founded, with help from British Abolitionists. Although the French were far less active than their British counterparts, the *Société* pressed for the abolition of the slave trade and, eventually and without compensation, of slavery. One of its founders, Jacques-Pierre Brissot (1754–93), argued that, with education, black people had the same capacities as white people. A journalist who had been jailed for criticizing the Queen, Marie-Antoinette, Brissot had visited the United States and inquired about the slave trade there. Having fallen foul of Maximilien Robespierre, the head of the Jacobins, Brissot, a one-time radical who now appeared moderate, was executed in 1793.

In the utopian idealism of the French Revolution, the liberties affirmed by the Revolutionaries were believed by many radicals to be inherent in humanity as a whole, to be human rights, and thus to be of global applicability. In January 1792, the attention of the National Assembly was directed by its Colonial Committee towards Madagascar, from which France had traded for slaves since the seventeenth century. Instead of a French territorial expansion to be achieved by conquest, there was a call:

not to invade a country or subjugate several savage nations, but to form a solid alliance, to establish friendly and mutually beneficial links with a new people ... today it is neither with the cross nor with the sword that we establish ourselves with new people. It is by respect for their rights and views that we will gain their heart; it is not by reducing them to slavery ... this will be a new form of conquest.[16]

Initially, the slave trade was not banned by Revolutionary France; indeed, it reached its peak during the years 1789–91. This reflected the value of the West Indies to the French economy. In this stage, when the Revolution was not radical, it was argued that slaves were not French and, therefore, that slavery and revolution were compatible, and the slave trade was normalized by a decree of the National Assembly in March 1790.[17]

The major slave rising in France's largest and most prosperous Caribbean colony, Saint-Domingue (modern Haiti), in 1791 transformed the situation and also helped competing sugar-producing areas: the British colonies and, especially, Brazil. This rising exploited the serious divisions between the white settlers that arose from differing political responses to the French Revolution. These divisions provided the slaves with opportunities greater than those that had confronted previous slave rebellions. The rising led to a complex conflict in Saint-Domingue in which, in 1793, the Civil Commissioner, Léger Sonthonax, freed the slaves in the Northern Province in order to win their support. The following year, the National Convention in Paris, the republican successor of the National Assembly, abolished slavery in all French colonies, a step that was unwelcome to most of the white settlers, as well as to the government of South Carolina. The abolition was less the result of ideological conviction than a recognition of a step already taken by their representatives in the colony. French Revolutionary leaders were divided over the issue, with Robespierre arguing that Georges Danton seriously harmed the French economy by the abolition, which he saw as a rash act.[18]

The abolition of slavery did not protect the French position in Saint-Domingue, which instead, after a bitter war, became the independent black state of Haiti. There were important African echoes in the rising and the subsequent warfare, but the rising also drew on European military practices and on the ideology of revolutionary France. Toussaint L'Ouverture, the leader of the rising, seized the Spanish side of the island (modern Dominican Republic) in 1800.

Under Napoleon, a major effort was made to restore the French position. This was not really a case of French Revolutionary idealism falling victim to a reaction because Napoleon's reinstitution of slavery in 1802 was in line with a key element of the pragmatic motivation of abolition in 1794. Slavery was restored in Guadeloupe and Martinique in 1802 by Napoleon and the resistance of former slaves on Guadeloupe was rapidly suppressed.

The entry to France of West Indian black and mixed-race people was pro-hibited.[19] The French sent a major force to Saint-Domingue under Napoleon's brother-in-law, Charles Leclerc, to reconquer the colony. This expedition benefited from mounting successful amphibious operations, while Toussaint was treacherously seized during negotiations.[20]

However, his successor, Jean Jacques Dessalines, drove out the French, who were hit hard by yellow fever, in 1803.[21] Fighting there demonstrated what was also seen with the British West Indies regiments, that black people were far better than Europeans as warriors in the Caribbean, especially because their resistance to malarial diseases was higher.[22] Prefiguring their very different later role in subsequently ending the slave trade, the British played an important part in ensuring the success of the Haitian revolution, with a crucial blockade of Saint-Domingue's ports in 1803 when war with France resumed. This wrecked the French attempt to recapture the colony. The French withdrew from the island with British cooperation, and, on January 1, 1804, the independence of Saint-Domingue as the new state of Haiti was proclaimed.

A black state seeking rapid abolition and a move beyond racial prejudice proved far too much for the influential slaveholding interests in America—it was perceived as a threat to the racial order there. Thomas Jefferson, America's President, a Virginian, and a slave owner, refused to recognize Haiti. It was not until most of the slave states in the American South had disenfranchised themselves by their rebellion in the Civil War (1861–5) that the independence of Haiti and Liberia were recognized: in April 1862, Congress authorized the dispatch of American envoys. Opposition to recog-nition then was led by Senator Garrett Davis of Kentucky, a representative of one of the border states—slave states that backed the Union. He claimed to be able to imagine no sight so dreadful as that of a "full-blooded" black person in Washington society.[23]

In a context of news and information flowing across the Caribbean and Atlantic by word of mouth, opening up prospects for new outcomes,[24] there were also slave revolts elsewhere in the Caribbean world in the 1790s, includ-ing, in the British colonies, Fedon's rebellion on Grenada in 1795–6, and risings or conspiracies in Dominica and St. Vincent, followed by a conspir-acy on Tobago in 1802. All were unsuccessful, as was the Pointée Coupée slave rebellion in Louisiana in 1795. However, the success of the Haiti revo-lution led to a climate of fear in white society, which was accentuated by the killing of settlers on Grenada in 1795. The Maroons (escaped slaves) of Dominica, who were already a serious problem in 1785, were only sup-pressed in 1814, in part as a result of defections and in part due to the burn-ing of their cultivated patches. In the Dutch Caribbean colony of Curaçao, there was a revolt in 1795, in which about 2,000 slaves and free black people rebelled only to be defeated by the local militia, some of whom were black. Slave rebels were less successful in Latin America than in Saint-Domingue. In Venezuela, a rising in 1795 was suppressed. The small-scale "Revolt of the Tailors" in Salvador in Brazil in 1798, which included slaves as well as mulattos and white people, called for the abolition of slavery, but was suppressed, as was a similar conspiracy in Pernambuco in Brazil in 1801. White fear, which was extensive, focused on concern about a "second Haiti."[25]

Britain

The British situation in the 1800s differed from that in France. There was a boom in sugar exports from the British West Indies in the 1790s caused by the chaos in Saint-Domingue. Moreover, widespread opposition to populist reform, let alone radicalism, that stemmed from hostility in Britain to the French Revolution led to a shadowing of Abolitionism. However, in contrast to France, there was then an upsurge in Abolitionism in the 1800s. This upsurge led to the formal end of the British slave trade.[26]

The reasons for the end both of this slave trade and of slavery itself has been a cause of much controversy, controversy that continues to the present. Some commentators focus on a lack of profitability caused by economic developments, rather than on humanitarianism on the part of the British.[27] This view, however, underplays the multiplicity of factors that contributed to Abolition. Economic problems in the plantation economy of the West Indies were significant. They stemmed from the impact of the American Revolution and subsequent protectionism on the trade between North America and the West Indies that was so important to the supplies for and markets of the latter.[28] Yet, there are indications that slave plantations in the West Indies remained profitable.[29] In part, this profitability reflected the ability of plantation owners to innovate. An aspect of this innovation included better care for the slaves. Aside from its continued profitability, the West Indies' plantation economy remained an important asset base. Furthermore, the limited convertibility of assets did not encourage disinvestments from slavery: too much money was tied up in mortgages and annuities that were difficult to liquidate in a hurry, and, at least in financial terms, the planters had a good case for the generous compensation they pressed for and received, rather than the loan originally proposed. The moral case was totally different.

Instead of problems within the British slave economy, it is more appropriate to look at the outside pressures towards Abolition. These pressures led, indeed, to a situation in which it became the general assumption of the "official mind" that action against the trade was a proper aspect of British policy.[30] Already, the lengthy and difficult war with France had led from the late 1790s into the early 1800s to a drive for reform designed to enable Britain to meet the challenge. This drive had included significant changes, including the introduction of income tax, the departure of sterling (the currency) from the gold standard, the extension of the detailed mapping of the country by the Ordnance Survey, legislation to regulate both the press and labor relations, parliamentary union with Ireland, and the establishment of a national census. The connection between these innovations and the abolition of the slave trade was indirect, but they were linked in a sense that significant changes were necessary. Thus, a conservative administration embraced reform and used the national emergency to sell reforms to its more conservative supporters.

Pressures toward Abolition included, and contributed to, a marginalization of groups, especially West Indian planters, that had encouraged and profited from British and European demand for tropical goods. In contrast, the reforming, liberal, middle-class culture that was becoming of growing importance in Britain, and ably that used the newspapers to articulate its views, regarded the slave trade and slavery as abhorrent, anachronistic, and associated with everything it deplored. Abolitionists indeed were encouraged and assisted by a confidence in public support, and this confidence helped influence the debate amid the elite. Moreover, all sorts of reform impulses in Britain converged on Abolitionism, including concern over the treatment of animals, which provided a context within which the brutal conditions to which slaves were exposed appeared doubly abhorrent. Thomas Burgess was also an early supporter of the RSPCA.[31]

Abolitionism, moreover, offered a country tired by the travails of a seemingly intractable war with Napoleon, the opportunity to sense itself as playing a key role in the advance of true liberty. Abolitionist medals show how self-conscious this sense was.[32] This attitude was particularly valuable in 1806, as Britain's allies succumbed after crushing French victories at Austerlitz (1805) and Jena (1806) and the Third Coalition of powers against France collapsed as a result, with first Austria and then Prussia surrendering to France. Indeed, the parlous state of the war encouraged the idea that the Abolition of the slave trade was necessary for Britain to benefit from divine providence.[33] This approach focused Evangelical ideas. The transformation of God's work in order to build a new Jerusalem in accordance with divine providence was an attractive view that aligned human inventiveness with moral purpose. Thus, eighteenth-century Enlightenment ideas were transmitted in a new language, although still with the underlying idea of a benevolent God who had equipped humans with the means to serve his purposes, a theme with a strong background in the Judeo-Christian tradition. There was the powerful conviction that moral purpose was necessary in all endeavors to assuage divine wrath, as with the ending of the slave trade. Providentialism was probably a more significant factor in leading to support for ending the slave trade than some of the secular calculations that are generally mentioned in modern discussion.

Banning the Trade

In 1805, the ministry was led by William Pitt the Younger, the First Lord of the Treasury from 1783 to 1801 and 1804 to 1806. He was a reformer by temperament and conviction, as well as a statesman who profited from his appeal to the reforming middle-class constituency, a constituency that included Wilberforce and many others opposed to slavery. Ministries lacked the relative coherence and relative party discipline of more modern British administrations, not least because, appointed by the Crown, ministers opposing their colleagues were sometimes able to rely on the monarch's encouragement. Pitt's ability to lead a ministry that included more conservative politicians was conditional. In 1785, he had failed, in the face of opposition by ministers encouraged by George III, to push through a reform in the franchise (right to vote), and, in 1801, he resigned when George opposed his support for Catholic emancipation, ensuring that Catholics had similar rights to Protestants. This political tension affected the ministry's consideration of the slave trade, both prior to Pitt's resignation in 1801 and after he returned to office in 1804. In 1805, the ministry issued Orders-in-Council (government decisions taken by the Privy Council) that banned the import of slaves into newly captured territories after 1807 and, in the meantime, limited the introduction of slaves to 30 percent of the number already there. This measure both reflected opposition to the slave trade and was presented on prudential grounds, as a way to limit the economic strength of these territories when some were returned as part of the peace settlement at the end of the war as they would be: Cuba, Guadeloupe, and Martinique had been returned by Britain in 1763, and the last two were to be returned anew after the Napoleonic War. Colonial gains had been returned, most recently in 1783, at the end of the War of American Independence, and in 1802, under the Treaty of Amiens.

These Orders-in-Council were taken much further by the next government, the more reformist, coalition Ministry of All the Talents, which took power after Pitt's death in January 1806. That year, the new ministry, in which William, Lord Grenville and Charles James Fox were the key minsters, supported the Foreign Slave Trade Act, ending the supply of slaves to conquered territories and foreign colonies. This reduced the value of slaves to slave traders as such sales had proved a way to raise the price of slaves, a measure that could anger plantation owners who sought low prices for slave purchase.

The highpoint of the Abolitionist process occurred when the Abolition Act of 1807 banned slave trading by British subjects and the import of slaves into the other colonies. The bill, which was strongly supported by the government and which cooperated with Wilberforce, was introduced in the House of Lords in January 1807 and carried by 100 to 36. Grenville moved the second reading of the Bill in the Lords on February 5, declaring that "justice" was the principal foundation of the measure, adding, "this trade is the most criminal that any country can be engaged in ... how much guilt has been incurred in carrying it on," and blamed war in Africa, and "barbarity" on its coast, "we have every reason to conclude that it is the temptations held out to the chiefs on the coast of Africa, for the gratification of their passions, that induces them to enter into those frequent wars."[34] The key debate in the Commons, that on February 23 saw, after a debate of ten hours, a division at about 4 a.m. on the 24th of 283 to 16. Wilberforce told the Commons that it was clear from the debate "that where a practice was found to prevail inconsistent with humanity and justice, no consideration of profit could reconcile them [the parliamentarians] to its continuance." The bill received the royal assent on March 25, 1807. Subsequently, in 1811, participation in the slave trade was made a felony, a measure that underlined opposition to the trade.

Pressing Powers to End the Trade

Britain also used its considerable international influence and strength to put pressure on other states to abolish or limit the slave trade. Not only did the trade now seem morally wrong, but, once abolished for British colonies, it was also seen as giving an advantage, real or potential, to rival plantation economies. This made the British imperial economy vulnerable, and thus potentially challenged the Abolitionist cause politically. Naval strength, amphibious capability, and transoceanic power projection ensured that the British were in a dominant position and very well placed to advance their views. Once war resumed with Napoleon in 1803—a war that lasted until 1814, and was briefly and successfully resumed and concluded in 1815—the British seized St. Lucia, Tobago, Demarara, Essequibo (now both in Guyana), and Surinam in 1803–4, successes that made the Orders-in-Council issued in 1805 significant. The British followed these gains with the Danish West Islands—St. Croix, St. Thomas, and St. Johns in 1807, with Martinique and Cayenne from the French in 1809, and with Guadeloupe, St. Eustatius, and St. Martin in 1810. Fort Louis at the mouth of the River Senegal, the last French base in Africa, fell in 1809. Having been returned after peace was negotiated in 1814, Martinique and Guadeloupe were seized anew in 1815 after Napoleon briefly returned to power in France.

Although their pressure was widely resented by others, both during the Napoleonic War and thereafter, being regarded as self-interested interference and undesirable moralizing, the British were in a position to make demands. In 1810, pressure was exerted on Portugal, then very much a vulnerable and dependent ally, its territories protected from Napoleon by British troops and warships, to restrict the slave trade as a preparation for Abolition. Because

Brazil and Angola were Portuguese colonies, the Portuguese position was particularly important. The trend of both war and peace favored British demands. Under British pressure, the Congress of Vienna, which sought to establish a post-war order, issued in February 1815 a declaration against the slave trade.[35] In 1815, Napoleon, on his return from Elba to challenge that order, abolished the French slave trade, possibly as a way to appeal to progressive British opinion. Subsequently, after Napoleon was defeated at Waterloo by Anglo-German forces in 1815, the second-time returned Bourbon regime of Louis XVIII in France was persuaded by Britain to ban the slave trade. This was of great concern to British Abolitionists, as the French slave trade, if it continued, was seen as an opportunity for British investment. With France occupied by foreign troops among whom the British were prominent, the situation was propitious for British expectations. In 1817, an Anglo-Portuguese treaty limited the slave trade in Brazil to south of the Equator, ending the supply of slaves from the Guinea coast, and an Anglo-Spanish treaty contained similar provisions: Spain had rejected such pressure in 1814. In 1814, with effect from 1818, the Dutch slave trade was abolished. Again, this reflected British influence, as the Netherlands was also a dependent ally.

Meanwhile, in 1807, in an act signed into law by President Thomas Jefferson on March 2, with effect from January 1, 1808, the slave trade was also banished by the United States, although there was scant attempt to enforce the ban. Orders-in-Council, however, issued on November 11, 1807 were used by the British to justify seizing American slavers. The Court of Appeals of 1810 accepted the argument of the barrister James Stephens, Wilberforce's brother-in-law and a member of the evangelical Clapham Sect who had become convinced of the horrors of slavery by his time in the West Indies. Stephens argued that slave trading was a violation of the law of nations, the laws of humanity, and British and American law, and that, therefore, neutral slave ships that had been captured could be legitimately seized.[36] America was then the leading neutral power.

During the War of 1812, the conflict of 1812–15 between Britain and America, the British willingness to receive and arm escaped slaves aroused American anger. British commentators suggested encouraging slave resistance as a way to weaken the USA.[37] Slaves and free black people served with both sides, part of a long practice in New World warfare of seeking military labor. The slaves wanted freedom, but promises were generally broken—for example, by Andrew Jackson when seeking support in 1814–15 for his defense of New Orleans against British attack.[38] About 4,000 slaves fled from the USA with the British at the end of the war. British naval commanders refused to return escaped slaves who had taken refuge with them, arguing that they were not property, the return of which was stipulated by the treaty. This stance led to eventually successful post-war American demands for compensation.

The Continuing Slave Trade

Despite British governmental pressure, the Atlantic slave trade continued, not least because slavery had not been abolished, either in the British colonies or in those of the other European powers. British participation in the slave trade persisted, both legally and illegally, directly and indirectly. This participation included the purchase of slaves, both for British colonies and for British-owned operations elsewhere. These operations included mining for gold in Brazil and for copper in Cuba, as well as the provision of goods, credit, insurance, and ships to foreign slave-traders, and direct roles in slavery.[39] Delays in ending the slave trade also encouraged the purchase and shipping of slaves that was designed to preempt the end of legal imports. The provision of investment and working capital was particularly valuable to slave societies, and this provision was an aspect of Britain's dominance of what was otherwise an undercapitalized Atlantic world. As also in free societies, British finance helped support rail construction, and this finance was important to plantation economies, in Brazil, Cuba, and the USA. The British were not alone in this process. American manufactured goods supplied to Africa came to play a significant role in the slave trade from the 1840s.

Although demand for slave labor was in large part met from the children of existing slaves, the continuation of slavery ensured that, even where the slave trade had been abolished, the smuggling of slaves continued. Moreover, in response to action against the slave trade, there was a search for new sources of slaves. This led to the development of southeast Africa as a source for slaves. The slaves were in part purchased in return for textiles shipped from the British base of Bombay (Mumbai). However, the distance of this source from the Americas hit the profitability of this trade, and as a result most slaves shipped via the Portuguese bases in Mozambique, such as the ports of Mozambique and Quelimane, went to nearby Mauritius and Réunion, and not to Brazil, the biggest market for slaves.

Demand kept the international slave trade alive. More particularly, the slave trade to the leading and, with the prohibition of many other traders, increasingly important market, Brazil, was not effectively ended until 1850, and that to the second market, Cuba, until 1867; slavery continued in both. Cuba, which, until conquered by the USA in 1898, was a Spanish possession, imported an annual average of 10,700 slaves in 1836–60 and the number of slaves there rose from 217,000 in 1810 to 437,000 in 1847, with the area of sugar cultivation greatly expanding.[40] The profitable nature of the sugar economies of Brazil and Cuba, the commitment to the lifestyle and ethos of slaveholding, and a lack of relevant European immigrant labor, kept the trade successful. Rising world sugar and coffee prices in the 1840s encouraged the demand for slaves. Economics was not the sole element. In Brazil and Cuba, slavery was also part of the political bargain underlying the state. In Brazil, the continuation of slavery encouraged landowners to remain loyal to the new independent state, rather than backing separatist movements which were very frequent there, notably in the 1830s. In Cuba, slavery encouraged the loyalty of the landowners to the Spanish colonial regime, but many of the slaves were harshly treated.[41]

The slave trade provided revenue for the European positions in Africa, whence the slaves were shipped. In particular, most of the annual revenue of Angola, which remained a Portuguese colony, derived from export duties on slaves shipped out to Brazil. The flow of slaves to Brazil was principally financed by the shipping to Angola, in return, of textiles. Cheap brandy and firearms were the other major goods shipped to Angola. Britain and Brazil were the leading sources of the textiles, and Britain's role was important to Anglo-Brazilian trade. The role of this trade helped complicate the attitudes of the British government to the continuation of the slave trade to Brazil. There was pressure on the government over the issue on behalf of British manufacturing interests.

Demand for slaves encouraged supply, and also shifts in the supply system. For example, the number of slaves shipped through Portuguese-controlled ports in the Cabinda region, to the north of the River Congo, rose in the 1820s. In part, this rise was due to the decline of the French and Dutch trade from this region but, in part, to the expansion of the Atlantic slave trade further east into the African interior.[42] In addition, until the late 1830s, the Bight of Benin and, until the 1850s, the Angolan coast north to Cabinda remained important sources of slaves.[43] The export of slaves to the expanding economy of Brazil rose in the 1840s. The British role in the Bight had been replaced by Brazilian, Dutch, Portuguese, and Spanish traders. Furthermore, American slavers greatly profited from supplying demand in Brazil and Cuba.

The raiding warfare that provided large numbers of slaves remained important across Africa[44] and reflected shifts in the Atlantic transit system. For example, Opubu the Great, ruler of the important port of Bonny on the Bight of Biafra (1792–1830), responded to British moves against the slave trade by selling palm oil to Britain while, at the same time, developing his slave interests with Portugal.[45] As a result of the continued flow of new slaves, Brazil and Cuba remained more African in the nineteenth century, and notably in the second half of the century, than the British West Indies or the southern USA. This had important long-term consequences in terms of their societies and cultures; consequences which continue to this day.

America

In America, the initial acceptance of slavery was a product of the federal character of the new state, and of the role of slave holding, not only in the economies of the Southern states, but also to their sense of identity and distinctiveness. When the Union was originally created, it had been agreed that abolition of the slave trade would take place, an abolition agreed in 1807, but there was no such provision for slavery. Moreover, the possibility of such a development appeared remote as a result of changes in the South. Much of the discussion and representation of slavery for the early years of the Republic related to the North African enslavement of American sailors from 1785 which led to conflict in 1801–5 and 1815. Thus, Susanna Rowson's play *Slaves in Algiers* was staged in Philadelphia in 1794.

After 1808, the slave trade to America was no longer legal, but, within that country, the extent of the slave states ensured that there was still a very extensive slave trade, particularly from the Old to the New South: from the Chesapeake to states further south and south-west. The consequences included an accentuation of the family breakup that was a characteristic and always a risk of slave life.[46] This situation was similar to that in Brazil, where the sugar planters of the North-East sold slaves to the coffee planters further south, who were expanding west into the province of São Paulo using the railway to create new links and opportunities. Similarly, delays in ending slavery provided a market for the slave trade within the Caribbean, particularly once direct trade with Africa was limited by British naval action. Caribbean slave supplies, for example, became more important to the Spanish colony of Puerto Rico from 1847.[47] Aside from slave sales, the prevalence of slave hiring in the American South ensured considerable geographical mobility among slaves. This helped keep slavery responsive to the market, and thus both an efficient economic system and, more generally, part of an effective economy.[48]

The slave economy in America was transformed as a result of the major expansion of cotton. This owed much to Eli Whitney's invention in 1793 of the cotton gin, a hand-operated machine which made it possible to separate the cotton seeds from the fiber. This possibility encouraged the cultivation of "upland" cotton. It was hardy, and therefore widely cultivable across the South, but was very difficult to deseed by hand, unlike the Sea Island cotton hitherto grown, which had been largely restricted to the Atlantic coastlands. Annual cotton output in the USA, as a result, rose from 3,000 bales in 1793 to over 3 million in the 1850s. American cotton became the key source for British cotton manufacturers, who were a major force in the British economy, and notably in the economic powerhouse of Manchester, the leading city in Lancashire, whose port was Liverpool. Their quest for profit was to be hit hard by their greedy response to the Union blockade during the Civil War.[49]

The profitability of the cotton economy was important to the continued appeal of slavery in the South, and, as tobacco became less well capitalized, so slaves from the tobacco country were sold for work on cotton plantations. This ensured that slaves became less important in the Chesapeake states of

Maryland and Virginia. The success of the cotton economy, and the ability to boost the birth rate of American slaves, were such that Southern apologists did not regard the slave system as anachronistic—indeed, far from it.

This belief, and the resilience of slavery, encouraged Abolitionists to press for the end, an enforced end, to slavery as well as of the slave trade. The resulting controversies were fought out across the political geography of America and what became America. In 1819, when the Missouri Territory applied for statehood, the proposals of James Tallmadge, a New York Congressman, for the gradual ending of slavery in the territory (by prohibiting the entry of new slaves and freeing all existing slaves born after admission to statehood once they turned 25), won extensive support in the North, but was seen in the South as a threat to its identity and existence. The subsequent Missouri Compromise was regarded as a challenge by Southerners opposed to Northern interference, and encouraged them to press for America's expansion to the west and also to develop an interest in Mexico and the Caribbean so that new slave-owning regions could be added to the country.

Texas was one area of opportunity. While Texas was under Mexican rule (1821–35), the attempt by the Mexican government to prevent the import of slaves there aroused much anger among the American colonists. Texas became independent as a country with slave ownership in 1836. British recognition of the then-independent Republic of Texas in 1840 was made on the basis of its prohibition of the slave trade. Charles Elliot, the British envoy in Texas, advanced the idea of an independent pro-British Texas partly reliant on free black labor, and thus a suitable ally for Britain. Instead, Texas abandoned independence and joined the USA as a slave state in 1845, thus strengthening the Southern bloc.

Whereas opposition to slavery was violently suppressed, notably Nat Turner's rebellion in Virginia in 1831,[50] attempts to extend slavery were seen in the mid-nineteenth century, with American filibusters trying to seize and hold territories by force in order to establish new slave societies. William Walker (1824–60) was the filibuster who made the most splash. A Tennessee-born doctor, Walker became a newspaper editor in San Francisco, a city of apparently boundless prospects. There, he conceived the idea of a Central American state that was to fulfill his views of a manifest destiny for white men and himself. He sought to create a state ruled by white people and resting on a slave-based agriculture, a transposed South. Walker's first step was an invasion of Lower California, part of Mexico, in 1853. Capturing the capital, La Paz, Walker declared that it was independent, as was the neighboring Mexican province of Sonora. He put the new "Republic of Lower California" under the laws of Louisiana, which provided a legal basis for slavery. The Mexican army swiftly drove Walker's band out. In 1855, Walker tried again, this time seeking to exploit the civil war in Nicaragua. Having conquered it, he declared himself president in 1856. The Emancipation Edict of 1824 forbidding slavery was annulled, and Walker encouraged the idea that Nicaragua would be a key partner of the South. The following year, Walker was defeated and left.

Britain and Latin America

Britain expended much diplomatic capital on moves against the slave trade, so that the granting of recognition to the states that arose from the collapse of Spain's empire in Latin America in the 1810s and 1820s depended on their abolition of the slave trade. British support was important to Abolitionism in formerly Spanish America, but so also was the example and process of rebellion against Spain which had led to independence. These rebellions challenged existing patterns of authority, but many slaves backed the Royalists in part because they had more reliance on protecting their interests through royal justice, and were wary of independence.[51]

Across much of Spanish America, but not, to the same extent, (Portuguese) Brazil, there was a breakdown of order far greater than that in the American War of Independence. This breakdown was exploited by many slaves in order to escape or rebel. Slavery itself was abolished in Chile in 1821 and in Mexico in 1829. In Argentina, Peru, and Venezuela, however, Abolition was a gradual process. Alongside the diversity of the situation within Latin America, there were similarities there, not least in contrast to the USA. In particular, from the sixteenth century, racial classifications had been relatively fluid in Latin America, with racial mixture and caste both playing a major role alongside race. Moreover, there was manumission (the freeing of slaves) before Abolition, many slaves won freedom by participation in the wars of independence, and, alongside many in misery, there were black elites.[52]

Pressure was exerted by Britain on other states, including France and the USA, to implement their bans on the trade, although there was considerable anger on their part about British demands, not least over the issue of searching ships. Indeed, British pressure was in part countered by the continuation of slavery in the USA, and American influence, in particular, helped in the continual slave trade to Cuba. So also did the lack of a Spanish Abolitionist movement. In 1839, the Palmerston Act authorized British warships to seize slave ships registered in Portugal and sailing under the Portuguese flag, a measure in part intended to hit the use of the flag by Brazilian slave dealers. Some traders then switched to the use of the French flag. New treaties to enforce the end of the trade were signed with Portugal in 1842 and with France in 1845.

This issue caused particular problems in Anglo-Brazilian relations. In 1826, Brazil, concerned about its international position, not least in terms of the complex politics involving Portugal, accepted a treaty with Britain, ratified in 1827, promising to make the trade illegal within three years of ratification. Furthermore, in 1831, the Brazilian General Assembly passed a law ordering the liberation of all slaves entering Brazil. In anticipation of Abolition, the treaty led to a marked rise in demand for slaves, and also in their price, in 1828–30. There was also renewed interest in the recruitment of Native Indian labor. Thereafter, demand for slaves and prices fell in 1830–3, before both, however, rose anew. This reflected the pathetic nature of enforcement in Brazil, which arose from a strong sense that slavery and the slave trade were essential, a sense that drew on the continuing demand for slaves, as well as from anger concerning British interference and about Brazilian measures to enforce the law. In the late 1830s, political pressure for the end of restrictions on the trade grew, and it was openly conducted. The Brazilian navy came to do very little against the trade. The low price of slaves in Africa encouraged the revival in the trade. The inflow of slaves in Brazil greatly increased in the late 1840s, so that by 1850 there were over 2 million slaves in Brazil.[53] Many worked in the booming coffee industry that benefited from greatly increased demand from the growing population of Europe, and specially the increasing wealth of an expanding European middle class, which enjoyed the disposable income to purchase an increasing amount of coffee.

In 1845, however, the British Parliament passed a Slave Trade Act authorizing the British navy to treat suspected slave ships as pirates. This led to the pursuit of ships into Brazilian waters, much to the anger of Brazil. Nevertheless, the Brazilians were not in a position to resist Britain, either politically or economically. Coffee and sugar could have been obtained by Britain from elsewhere, and Brazil needed British capital. Greater British pressure was exerted from 1850 when the slave trade was formally abolished by Brazil in the Eusébio de Queiroz law, a measure that, in turn, owed much to British action. The British subsidized Brazilian Abolitionism, which was the lurid presentation of the fate of Saint-Domingue. More generally, by helping push up the price of slaves, which rose greatly in the 1850s and yet more thereafter, British pressure ensured that slave owning was too expensive for many Brazilians, which reduced its role.[54] As a result, alongside changes within Brazil, the end of the South Atlantic slave trading system owed much to the power of the North Atlantic.

Pressure was also exerted on the Spanish colony of Cuba, sufficiently so for David Turnbull, the British Consul in 1840–2, to be accused of incitement to slave risings.[55] Conversely, the slaveholders who dominated Cuban society looked for encouragement and support to American Southerners. The retention and abolition of the slave trade and slavery have to be set in international contexts, both political and economic. Indeed, the history of slavery, as of the slave trade, is thus a central part of international history.

The British Navy Attacks the Slave Trade

The sense of moral purpose behind British policy rested on the state's unchallenged naval power,[56] and was given a powerful naval dimension by the anti-slavery patrols off Africa and Brazil, and in the West Indies. Even in 1807, when Britain was at war with France, and naval resources were very stretched blockading its ports, two warships were sent to African waters to begin the campaign against the slave trade. Indeed, Abolitionists pressed for the retention of British bases in West Africa that they could help in action against the trade.

The large-scale employment of naval pressure against the Barbary states of North Africa (Algiers, Tunis, Tripoli), which seized Westerners as slaves, acted as a bridge that helped to make such pressure against Western traders elsewhere seem more acceptable. In 1816, in the biggest deployment of this type, Admiral Lord Exmouth and a large British fleet, with the support of a Dutch frigate squadron, demanded the end of Christian slavery in Algiers. When no answer was returned, he opened fire: 40,000 roundshot and shells destroyed the Algerine ships and much of the city, the dey (ruler) yielded, and 1,211 slaves, mostly from Spain and the Italian principalities, were freed. This was seen as a great triumph. Exmouth was made a viscount, voted the freedom of the City of London, and was granted membership of chivalric orders in Naples, Sardinia, Spain, and the Netherlands. The British presented themselves as acting on behalf of the civilized world, and as assuming a responsibility formerly undertaken by the Bourbon powers (France and Spain). In 1819, a British squadron returned anew to Algiers, while, in 1824, the threat of bombardment led the dey to capitulate again to British demands. That year also, the bey (ruler) of Tunis was made to stop the sale of Christian slaves.

Like other North African centers, Tunis compensated for the decline in its supply of Mediterranean slaves as a consequence of successful action against raiding by increasingly tapping the trans-Saharan slave trade. In the 1790s, the Souk el Berka, the slave market of Tunis, sold as many as 6,000 slaves from sub-Saharan Africa a year. Similarly, visiting Tozeur, a major trading center in southern Tunisia, in 1727, Thomas Shaw, the chaplain to the British merchants at Algiers, noticed a "great traffick" in slaves. However, in the nineteenth century, this trade declined. As a demonstration of the range of factors that might be involved, competition from Tripoli, taxation in Tunis, and large-scale conflict in the *sahel* all played a role in hitting the trade.

In 1834, another outgunned British ship, the HMS brigantine *Buzzard*, under Lieutenant Anthony William Milward, took on the well-armed and larger Spanish brig *Formidable* off West Africa after a chase of seven hours. In a "smart action" of 45 minutes, the *Buzzard* had several injured and, as Milward reported "our fore and maintop-mast stays were cut, running rigging and sails much damaged, flying jib-boom shot away, and bumpkin carried away in boarding," but six of the slaver's crew were killed. 700 slaves were freed.

The most important active British anti-slavery naval force in the first half of the century was that based in West Africa (until 1840 part of the Cape Command), which freed slaves and took them to Freetown in Sierra Leone, a British colony for free black people where close to 100,000 enslaved people were resettled between 1808 and 1863, creating a large, cosmopolitan population. Despite very heavy losses from yellow fever, including a quarter of the force in the exceptionally bad year of 1829, the anti-slavery commitment led to a major expansion of this force from the 1820s to the 1840s. In the late 1830s, British naval action helped greatly to reduce the flow of slaves from the Bight of Biafra.[57]

Warships based in Cape Town, a British possession from 1806, also played an important role, and anti-slavery patrols were extended south of the Equator in 1839, enabling Britain to enforce the outlawing of the slave trade Brazil had promised in 1826, but failed to implement. In 1839, unilateral action was taken against Portuguese slavers after negotiations had failed. The achievements of the warships were celebrated in Britain, so that the capture of the slave schooner *Bolodora* by HMS *Pickle* in 1829 led to a painting by William Huggins that was engraved by Edward Duncan.

The *Voladora* was larger and had a crew twice the size, but the *Pickle* under J.B.B. MacHardy closed, and after an action of 80 minutes the *Voladora*, the mainmast shot away, the sails repeatedly holed, and rigging trailing over the stern, surrendered. The British had lost four men, their opponents at least fourteen. 223 African men and 97 African women who had been bought in Africa were freed. 32 slaves had already died on the voyage. The British crew imprisoned the slavers in their own chains.

The *Pickle* was not alone. Five days after its victory the navy's smallest warship, the schooner HMS *Monkey*, under Lieutenant Joseph Sherer, captured the far larger Spanish brig *Midas* after an action of 35 minutes even though the *Monkey* had only one 12-pounder cannon and a crew of 26, while the *Midas* had four 18-pounders, and four 12-pounders, and a crew of over 50. *Midas* had bought 562 slaves from Africa, but only 369 were still alive when she was captured. Earlier in 1829, the *Monkey* had already captured an American slaver and a Spanish one, the latter, again more heavily gunned, carrying 206 enslaved slaves.

Anti-slaving activities were not restricted to the Atlantic, but were also important in the Indian Ocean and in East Asian waters. Thus, in 1821–3, the frigate HMS *Menai*, based at Mauritius, took action against slavers. There were also operations against piracy, which was often focused on slave-raiding—for example, off Sarawak in northern Borneo in 1843–9.

The advent of steam power added a new dimension to the naval struggle. It increased the maneuverability of ships, making it easier to sound inshore and hazardous waters, and to attack ships in anchorages. This made a major difference in the struggle against the slave trade, as slavers were fast, maneuverable, and difficult to capture, and could take shelter in inshore waters. It was also necessary for the British navy, from the 1840s, to respond to the use of steamships by slavers keen to outpace the patrols—another instance of the adaptability of the slave trade and slave economy, and its openness to new technology and investment. In West Africa, Lagos, a major slaving port fed with slaves by the serious Yoruba civil wars, was attacked in 1851, with a British steamship playing a prominent role in the attack, and the slaving facilities were destroyed. Lagos was annexed by Britain a decade later.

As part of the attack on the slave trade, there was the hope that the economic potential of West Africa could be harnessed in order, through the use of free labor, to provide an effective complement for the British Empire. Macgregor Laird, who founded the African Steamship Company, sought in the 1830s with the use of steamships to make the River Niger in West Africa a commercial thoroughfare for British trade, which he hoped would undermine the slave trade. This was also a theme of Sir Thomas Buxton's *The African Slave Trade and its Remedy* (1839), and of the Society for the Extinction of the Slave Trade and the Civilization of Africa. The political, economic, and, in particular, environmental difficulties facing this attempt led to its failure, but, alongside the participation of several with a slavery past, these ideas reflected the role of Abolitionism in strategies of betterment, both for Africa and for the West. The Niger expedition of 1841 fell foul of malaria.

The role of the British navy ensured that opposition to the slave trade would not simply be a matter of diplomatic pressure. It also meant that there was a constant flow of dramatic news for the press to help keep Abolitionism at the forefront of attention in Britain. The role of the navy also demonstrated the extent to which exogenous (external) pressures were crucial to the end of the slave trade.

Alongside the teleological tone of much writing about action against the slave trade, it is also necessary to note problems and criticisms. On July 12, 2006, the *Times* reported on the recently discovered journal of the young Lieutenant Gilbert Elliott, the son of the Dean of Bristol, who was a naval officer on HMS *Sampson*, serving on the slave patrol. He argued that the blockade was ineffective and that therefore it would be better to permit the trade to resume: "I am one of those who believe that while there is a demand there will be a supply, and that nothing will stop the trade unless we ruin the slave powers." At the same time, this was scarcely the fault of the navy, and Elliott's account of Africans who were freed underlined the plight brought or exacerbated by enslavement, including malnutrition, ill health, and exhaustion.

The American navy also took part in the struggle with slavery, sometimes in cooperation with the British off West Africa on the eve of the Civil War. This action overlapped with the protection of trade against privateering and piracy. The combined goals led to a major American naval commitment to the Caribbean from the 1820s, with operations offshore and ashore Cuba, Puerto Rico, Santo Domingo, and the Yucatán. These operations in part reflected concern that the Spanish colonial authorities were not being cooperative. In 1822, Commodore James Biddle commanded a squadron of 14 American ships in the Caribbean, and in 1823 David Farragut won notice in command of a shore party in Cuba while on anti-slavery duties. American naval activity also ranged further afield. In 1843, sailors and marines from four American warships landed on the Ivory Coast in West Africa in order to discourage the slave trade and to act against those who had attacked American shipping.[58] The Dutch navy, in contrast, made scant effort against slave traders. Once, under the treaty with Brazil ratified in 1827, Brazil had prohibited the trade in 1830, Brazilian warships played a role, although it met with scant government support and was soon restricted.

Pressure on Africa

Aside from British action against the slave trade itself, pressure was brought to bear on African rulers in order to agree to end the trade and, instead, to agree to legitimate trade.[59] This was an aspect of a more general interest in deriving benefit from inland Africa. For example, in 1812, Major-General Charles Stevenson sent Robert, 2nd Earl of Liverpool, Secretary for War and the Colonies, a memorandum urging the need to gain control of the city of Timbuktu on the River Niger:

> Africa presents a new country and new channels for your industry and commerce, its soils favourable for your West India productions, it produces gums, drugs, cotton, indigo … gold … iron … this to England is infinitely of more consequence than the emancipation of South America … the teak wood so famous for ship building might be cultivated with success in some of its various soils … the possession of Timbuktu would secure you the commerce of this quarter of the world and give you a strong check upon the Moorish powers of the Mediterranean by being able to intercept all their caravans and refusing them the commerce of the interior. It would likewise give you a complete knowledge of Africa to the borders of the Red Sea and to Ethiopia … at the same time you could raise black armies for your East Indies … not destined to conquer Africa, but bridle it, in order to have a check upon its kings, to protect British commerce as well as the African in its transit through the different kingdoms, by which means we should hold the country in check without the expense of defending it and by good management make the greater part of its sovereigns our friends, by supporting some, protecting others and augmenting their powers, and, as allies, drawing from them whatever black battalions we may want.[60]

As with the continuing slave trade, pressure was exerted in Africa within a context in which due allowance had to be made for the continued strength of its rulers. This was brought home in 1821, when the 5,000-strong British Royal African Colonial Corps under Colonel Sir Charles Maccarthy, Governor of Sierra Leone, was destroyed by a larger, more enthusiastic, and well-equipped Asante army. The governor's head, which became a war trophy and was used as a ceremonial drinking-cup, was a particularly lurid instance of Western failure. Macarthy's replacement, Major-General Charles Turner, recommended total withdrawal from the Gold Coast, but, instead, the British partly withdrew, so as to hold only the major coastal positions of Cape Coast Castle and Accra.[61] The Company of Merchants Trading to Africa was abolished that year and their bases were transferred to the British government.

New efforts and methods did not necessarily bring success in increasing Western control in Africa. For example, it proved difficult to use Africa's rivers. In 1816, an expedition sought to travel up the River Congo in order to discover if it was the outlet of the River Niger. Led by Commodore James Tuckey in command of the *Congo*, the first steamship on an African river, the expedition was blocked by difficult cataracts on the river, the boat did not operate correctly, and Tuckey and many of his men died of disease. No further progress was made until 1877 when Henry Stanley completed the first descent of the River Congo to the sea.

Indeed, in the 1810s and 1820s, Egyptian expansionism in north-east Africa was more obviously successful than its European counterpart in West Africa. This Egyptian expansionism continued to be important into the 1870s, with Darfur, Equatoria, and Harrar all acquired that decade. Slave raiding and trading, notably in Sudan and the Horn of Africa, were aspects of this spreading Egyptian control, a control that also had a religious dimension: Egyptian Muslims considered themselves entitled to enslave non-Muslims.[62] It was only from the 1840s that European power really became more insistent on the West African coast. French imperialism was extended, with new colonies established in Gabon (1842) and Ivory Coast (1843), while Spain established another colony, Rio Muni, the basis of the modern state of Equatorial Guinea, in 1843.

The impact of Western power could also be indirect. A prime instance was Tunis, which was newly vulnerable to Western power as a consequence of the French conquest of neighboring Algiers in 1830, with subsequent French expansion in what is now Algeria as well as pressure on Morocco in 1844. In 1846, Ahmed Bey of Tunis (r. 1837–56) made a state visit to France. Returning to Tunis, he sought to incorporate what he had seen in France and, in seeking a reputation as an enlightened ruler, abolished the slave trade and closed the slave markets.

Paradoxically, the Western slave trade was ended at the very time when the ability of Western states to project power into the African interior became stronger. This prefigured the later, very different, case that Western colonialism receded after World War II, while at a time when absolute Western military power reached a hitherto unprecedented level. As a similar point about chronology and the difficulty of using it to make causal points, slavery also came to an end when attitudes of racial superiority and Social Darwinism were becoming more clearly articulated. At the same time as the Western slave trade ended, enslavement continued in Africa as an aspect of warfare there—for example, in southern Africa, in what is now Botswana, the expanding Ngwato kingdom conquered and enslaved the Sarwa.[63]

The Fate of Slavery

It is very easy to move from the abolition of the Western slave trade to that of slavery, but it is important to note that these were not simultaneous and that there were cross-currents. In pressing in 1792 and 1807 for the abolition of the slave trade, William Wilberforce had denied that he supported immediate emancipation as he considered the slaves not yet ready, an argument later employed in Europe in the mid-twentieth century as the end of imperial rule and the pace of colonial independence were debated. Abolitionists had hoped that the end of the slave trade would lead to greater care of the remaining slaves by their owners and to the withering of slavery.

In contrast, however, there was soon a sense that the slave world was being strengthened at the same time that the slave trade was being ended. This was true not only of Mauritius in the Indian Ocean, which the British gained from France in 1810, but also of the colonies of Demerara-Essequibo and Berbice on the Guiana coast of South America, seized by the British from the Dutch in 1803. Plantation agriculture, the large-scale importation of African slaves, and a switch from cotton and coffee to sugar, all followed British conquest there.[64] They also did on Trinidad, seized from Spain by Britain in 1797. Thus, these colonies, all of which were retained by Britain under the Congress of Vienna peace settlement of 1814–15, were more like those of the late seventeenth-century West Indies than the more mature slave societies of the West Indies of the period, where a lower percentage of the slaves were African-born and where the work regime was less cruel. In the more mature slave societies, medical facilities for slaves were not significantly different from those for industrial workers in Britain.[65] Better care helped ensure that the British colonies were approaching demographic self-sufficiency by the 1820s, and thus less in need of obtaining new slaves by means of the slave trade. This self-sufficiency matched the situation in America, and underlines the extent to which the end of the trade did not necessarily weaken slavery. At the same time, in the West Indies the cost of acquiring and sustaining the workforce rose, in part due to moves against the slave trade and in part due to market forces. By the time of emancipation in the 1830s, the material consumption levels of the slaves were similar to those of manual workers in Britain.[66] Looked at differently, both groups faced a bleak prospect.

The continued strength of slavery helped lead to fresh Abolitionist pressure in Britain from the mid-1820s. In 1823, the House of Commons passed a resolution for the gradual abolition of slavery, although it had been modified to take more note of the slave owners' interests. This approach was an aspect of the emphasis on gradual emancipation as part of a managed close rather than either the defense of slavery as a positive good or the radicalism attributed to abolitionism.[67] In the West Indies, contention was more urgent. Some slaves believed that their owners were withholding concessions granted by the Crown. Slave owners, in turn, showed no desire to end slavery. Indeed, in 1830–1, in Jamaica there was talk of secession from British rule in response to Abolitionist pressure in Britain and to legislation aimed at the owners' powers of discipline over their slaves. Racism remained strong in the Caribbean world and was brutally displayed, not only in the commonplace callousness of the everyday treatment of slaves, but also in the harsh suppression of slave rebellions. Prominent episodes included rebellions in Barbados in 1816 (Bussa's rebellion), in Demerara (Guyana) in 1823, and in Jamaica in 1831–2, the Baptist War, the last in part in response to pro-slavery agitation among the white population. This was the largest slave rising in the British West Indies.[68] There were also slave risings in Virginia in 1800 and 1831, Louisiana in 1811, and in South Carolina in 1822, the Denmark Vesey conspiracy. This included a plan for the seizure or destruction of Charleston, although evidence for the plans has been questioned on the grounds that they were devised to give credence to the idea of a slave revolt. Religious zeal as well as ethnic consciousness played a role in some slave risings in this period—for example, on Jamaica and around Bahia in Brazil between 1808 and 1835, the *Malê* revolt in Bahia; while these risings also in part drew on the nature of warfare in West Africa.[69]

Action against slave independence was multifaceted. The American determination to end slave flight from Georgia to Florida lay behind the Seminole Wars (1817–18, 1835–42, 1855–8), as the Seminole Native Americans in Florida provided refuge for escaped slaves.[70] Indeed, in the second war, an armistice came to an end and Seminole resistance revived in 1837 when the Americans allowed slavers to enter Florida and seize Seminole and black people. In contrast, an important success for the Americans was obtained in 1838 when Major-General Thomas Jesup announced that black people who abandoned the Seminole and joined the Americans would become free, costing the Seminole 400 black fighters.

Opposition to slavery and to the conditions of slaves was also expressed in murders, flight, and suicide, and each was frequent. The conditions of slave labor in Brazil, especially but not only in the north-east, remained harsh and often violent, particularly in the sugar and coffee plantations. Food and clothing for the slaves was inadequate, and the work was remorseless, hard, and long. Death rates among slaves were high, in part due to epidemic diseases, but in part due to the work regime which was not abated in harsh weather. The conditions of work for pregnant slave women led to many still-births, while slave mothers lacked sufficient milk for their children. However, the 1872 census showed that 30 percent of Brazilian slaves worked in towns, and conditions were better there. A more humane treatment of rural slaves only began in about 1870 when their price rose.[71] The end of the slave trade benefited existing slaves.

In Cuba, which, like Brazil, was a low-cost producer, slavery remained important to the sugar monoculture of much of the economy, especially in western Cuba. The sugar economy depended on American investment, markets, and technology, while the British embrace of free trade helped Cuban production by ending the preferential measures that had helped ensure markets for sugar from Britain's colonies. Indeed, some British plantation owners emigrated from the British West Indies to Cuba. As a result, British belief in free trade encouraged pressure in Britain for the end of slavery everywhere, so as to create a level playing-field for economic activity. Moreover, the idea of free trade was linked to that of free labor.

Slavery Ends in the British Empire

In Britain, Abolitionist tactics against slavery reprised those earlier directed against the slave trade, with press agitation, public meetings, and pressure on Parliament. Concern about the plight of Christian slaves made the issue more urgent. Visions of Christian liberation from sin could be linked to, or symbolized by, the slaves breaking free from chains. Decreased confidence that the end of the slave trade would lead to the end of slavery was also significant. Pressure for immediate emancipation rose. Reports of the slave rising in Jamaica in 1831–2, and of the brutality of its suppression, helped make slavery appear uncivilized, and made action to stop it seem necessary and urgently so.

The Whig ministry of Charles, 2nd Earl Grey, that pushed through the Great (later First) Reform Act of 1832 that fundamentally revised the electoral franchise (right to vote) to the benefit of the middle class, also passed the Emancipation Act of 1833, which received the royal assent on August 28. Slaves were emancipated from August 1, 1834, Emancipation Day. Many Whig candidates had included an anti-slavery platform in their electoral addresses and Whig victories in the general elections of 1830 and 1831 were crucial. Thus, the furor over parliamentary reform was instrumental in the end of slavery. What, in the rhetoric of the time, might seem an illiberal situation in Britain was to be overthrown, followed by the same in the colonies. The Great Reform Act of 1832 contributed directly to the legislation as many seats with small electorates traditionally occupied by members of the West Indies interest were abolished and, in their place, came constituencies that favored Abolition. The latter seats tended to be large or medium-sized industrial or shipping towns, especially those with Nonconformists.[72]

As with the Great Reform Act and the franchise, there was an element of compromise in the emancipation legislation. Much of the parliamentary debate revolved around the financial issue of compensation. In the final settlement, the latter was raised to a grant of £20 million, which strengthened the planters' position in the West Indies. Moreover, compensation records indicate the wide-ranging impact of slave wealth in Britain, including the families, trustees, creditors and beneficiaries of slaveowners. The last spread this impact among many philanthropic causes and was a valuable source of liquidity.[73]

Initially, as a transition, all slaves aged over six were to become apprenticed laborers, obliged to work for their former masters for 45 hours a week—field workers for six years and others for four years. A clause, however, forbade the punishment of former slaves. In the end, this interim system, which led to protests from many former slaves, was ended in August 1838.

This, the first, lasting, end of slavery in a major state, indeed the leading empire in the world, was of great significance in global history. In immediate terms, the end of slavery indicated the extent to which great events do not necessarily have great causes in the sense of traumatic transformations. This end was not an episode akin to the fall of Rome in 410 or the launching of the Protestant Reformation in 1517. Moreover, as with many pieces of legislation, it is possible to draw attention to aspects that appear distasteful today. In particular, the issue of compensation strikes a wrong, not to say abhorrent, note. There is also the fact that the abolition of slavery in the colonies, like the Factory Act (1833) and the Poor Law Amendment Act (1834), followed the reform of the franchise in 1832, which was seen in Britain by most commentators as more important. The interim system that lasted until 1838, moreover, scarcely matches a triumphant account of the legislation. Yet, it is important to note that qualifications related to the other legislation as well. Moreover, this other legislation throws light on the harsh conditions of life in Britain. These conditions do not extenuate slavery but, rather, provide a broader picture of misery and, in particular, of the pressures of the world of work and the conditions that workers faced. The Factory Acts, which focused on conditions of employment in the British textile industry and did not extend to the colonies, still left work both long and arduous. The 1833 Act established a factory inspectorate and prohibited the employment of the under-nines. However, 9–10-year-olds could still work 8-hour days. In another instance of an interim system, by 1836 this provision would also apply to those under 13. Under-17-year-olds could work 12 hours. The 1844 Act cut the hours of under-13s to 6½ hours, and of 18-year-olds and all women to 12; those of 1847 and 1850 reduced the hours of women and under-18s to 10 hours.

So also with the Poor Law Amendment Act (1834). It introduced national guidelines for the case of the destitute in Britain, but the uniform workhouse system that it sought to create was not generous to its inmates. The able-bodied poor were obliged to enter workhouses, where they were to be treated no better than the conditions that could be expected outside, in order to deter all bar the very poor from being a cost to the community. Marriage and parenthood by the very poor were to be discouraged. In general, discipline was harsh. Thus, in the workhouse in Wimborne, meat was only provided once a week, while men and women were segregated. The condition of the British poor provided a field for missionary activity comparable to that which had earlier helped spur the abolition of slavery.

The end of slavery in its colonies was far from the end of the story as far as Britain was concerned. The already strong British opposition to the slave trade elsewhere was joined by action directed against slavery in other countries, as well as against British participation in slave economies—for example, in the Brazilian mining industry. In America, pro-slavery Southerners, especially from 1843, saw Britain as having ruined its West Indian colonies by emancipation, and as now intent on wrecking the competitive economies, notably the South, by making them abolish slavery. This was a less than complete account of British motivation.

The stress in Britain, in contrast, was on Abolition as a religious duty and a liberal necessity, while free labor was presented as appropriate for a society now adopting free trade as a moral as well as an economic good. The determination to reach out to all people was shown at the Great Exhibition in London in 1851, a totemic celebration of the potential of technology. There were displays by the Religious Tract Society and the British and Foreign Bible society of religious works in many languages, including the Bible in 165 languages. In Britain, Christianity was identified with human progress and was now seen as a key reason for supporting Abolition. Although Britain, like other Western societies, became more racist in the period, opposition to slavery was significant in its culture and was frequently expressed. British newspapers regularly carried notices of meetings against slavery—for example, in *Bell's Weekly Messenger*, on September 3, 1853: "On Monday Mr W. Brown delivered a lecture on American slavery before a large and respectable audience at the Lecture Hall, Greenwich" in London. Children's literature was also imbued with humanitarian sentiment.[74]

Slavery Ends in France

Anti-slavery was less important and popular in most of Continental Europe than in Britain, whether in Catholic France, or the Protestant Netherlands, in part due to the lack of a public politics comparable to Britain.[75] As in Britain, domestic political change played a key role. In France, in the more liberal July Monarchy that followed the 1830 revolution overthrowing the conservative Charles X, laws were passed in 1833 ending the branding and mutilation of slaves, and giving free black people political and civil rights. Effective action against the French slave trade was taken, in part because the new government, which sought good diplomatic relations with Britain, was more ready to ignore popular complaints about British pressure.

The end of slavery in the French colonies followed in 1848. The British example was significant in weakening French slavery, not least because of the prudential factor: British emancipation provided French slaves with new opportunities for escape to British colonies. In addition, the increased influence of reforming middle-class circles was important in France, although the decisive pressure came from a small group of writers and politicians, especially Victor Schoekher, who argued for immediate emancipation, rather than the slow processes that had long been favored. He became Colonial Secretary in the government that took power after a republican revolution overthrew the July Monarchy. As an instance of the slow processes formerly favored, the Mackau laws passed in 1845 had been restricted to helping slaves towards self-purchase. There have been efforts to explain the end of slavery as a response to slave unrest in the French colonies, but this explanation diverts attention from the crucial metropolitan context of decision-making. Nevertheless, major uprisings in the French Caribbean islands of Martinique and Guadeloupe in 1848 certainly speeded the application of emancipation, and the earlier argument that revolt was a possibility had been pushed hard by French Abolitionists seeking to convince opinion in France.[76] As in Britain and Denmark, there was compensation for the slave owners.

The end of slavery in the French Empire was highly significant, because, after Britain, France was the second most powerful overseas empire. With the loss of Haiti, Louisiana, St. Lucia, and Tobago in the early 1800s, the French had few remaining territories in the New World. However, France was becoming more significant as an African power, had a major fleet, and had become increasingly assertive on the world stage in the 1840s. Moreover, the end of slavery in the French Empire made it clear that British action would not be a one-off, but instead was beginning a trend that would influence major and minor powers alike.

In 1848, slavery in the Danish West Indian islands (now the American Virgin Islands) was abolished, when the threat of rebellion among the slave population forced the Danish Governor-General to free the slaves. Sweden had done so the previous year for its Caribbean colony of St. Barthélemy, while much of formerly Spanish America abolished slavery in the 1840s and 1850s, finishing the work already done there. The Dutch colony of Surinam followed in 1863, the USA in 1865, Spain in its colony of Cuba in 1886 (emancipation gradually began in 1870), and, most importantly, Brazil in 1888.

Slavery Ends in America

The profitability of the cotton economy in the South ensured that slavery in the America was not stopped by economic factors. Rather than slavery proving unprofitable or inconsistent with industrialization, or, indeed, challenged by slave revolts, political developments were crucial. Slavery, indeed, played a more prominent role in divisions within America from the 1820s, helping both to cement a sense of separate and particular Southern identity, and to give it an expansionist dynamic. Racial exclusion was presented as both form and focus of Southern cultural identity and defended on grounds of rights, necessity and prudence.[77] There was scant ebbing of support on the earlier pattern of Britain. At the same time, there were significant differences in the world of slavery, differences that greatly affected the pattern of behavior during the Civil War. The importance of the loyal border states (Delaware, Maryland, Kentucky, and Missouri) was such that even on January 1, 1863, when President Abraham Lincoln declared that Union victory in the Civil War would lead to the end of slavery, this related only to the Confederacy, and not to these states.

The contingent nature, both of American politics in the late 1850s and early 1860s, and of the course of the Civil War (1861–5), was crucial to the end of slavery.[78] At the same time, this contingency was given context by the development of radical public opinions, each based on a moral sentiment, one against and one for slavery. The admission of California as a free state in 1850 gave the free states a majority in the Senate, and the minority status of the South in the Union was a key feature of the sectional controversy of the 1850s, a feature that created serious problems for the South. Minnesota and Oregon followed as free states in 1858 and 1859 respectively. The volatility of the slavery issue was played out in politics, the law, and violence in the 1850s. On March 6, 1857, in a case centering on the effort by the slave Dred Scott to secure his freedom after the death of his owner, the Supreme Court decided as follows:

> It becomes necessary to determine who were citizens of the several States when the Constitution was adopted ... the legislation and histories of the times, and the language used in the Declaration of Independence, show, that neither the class of persons who had been imported as slaves, nor

their descendants, whether they had become free or not, were then acknowledged as a part of the people, nor intended to be included in the general words used in that memorable instrument.

However, the political tide was moving against this conservative view. A sense of being under challenge ensured that Southern secession was frequently threatened in the 1850s. In 1859, Richard, 2nd Lord Lyons, the British envoy, wrote: "after making due allowance for the tendency to consider the "present" crisis as always the most serious that has ever occurred, I am inclined to think that North and South have never been so near a breach."[79] Against the background of fears of slave insurrection that drew on the Haitian Revolution,[80] the alternative to secession was to seek to make the Union safe for the South and slavery, in part by reeducating Northerners about the constitution, or by acquiring more slave states, or by somehow addressing the vulnerabilities of the slave system in the South. The Southerners' failure to do so was compounded by the difficulties posed by the slaves' desire for freedom, although, in the context of a more oppressive white society in the South, slave flight did not have a scale nor disruptive consequences comparable to the situation in Brazil in the 1880s.

The 1860 election gave victory to Abraham Lincoln of the Republican Party, who wished to prevent the extension of slavery into the federal Territories. This was understood by Lincoln and others as a step that threatened Southern interests and identity. The election of Lincoln reflected the refashioning of politics by the slavery issue. This refashioning had led, first, to pressure on the Whigs, as with the rise of Libertyites and the Free Soil Party; subsequently, to the disintegration of the Whigs in the aftermath of the Kansas–Nebraska Act of 1854, and the related rise of the Republicans as a Northern sectional party focused on the restriction of slavery; and then to the division of the Democratic Party between Northern and Southern wings. In 1856, James Buchanan, the Democratic candidate, had won the South, but also Pennsylvania, Illinois, and Indiana, showing a winning national appeal; but in 1860, Stephen Douglas, the Northern Democrat, competed with John Breckinridge, the Southern Democrat. The Democrats split over whether there should be federal protection of slavery in the Territories, as the South wanted, or whether the decision should be left to the people in the Territories to decide for themselves. Southern politicians feared that the latter would lead to the erosion of slavery.

This competition between the Democratic candidates allowed Lincoln, the Republican, who carried the Northern states but none of the Southern ones, to win on fewer than 40 percent of the votes cast. National politics were no longer being contested by effective national parties, and, partly as a result, American mass democracy could not generate a consensus. Compromise was on offer, but no longer seemed sufficiently acceptable in the North and South to gather impetus. Lincoln rejected the proposal by Senator John Crittenden of Kentucky that the 36°30′ line of the 1820 Missouri compromise accepting slavery for the Arkansas Territory be run toward the Pacific, a line that would include the New Mexico Territory in the world of slavery.

Lincoln's election led to the secession of the South, beginning with South Carolina on December 20, 1860, and to the formation of the Confederate States of America as a body to rival the Union. Lincoln was willing to back a constitutional amendment prohibiting the federal government from inter-fering with slavery, but secession was unacceptable to him and the Republicans. They argued both that the maintenance of the Union was essential to the purpose of America as well as to its strength, and that it was necessary to understand that the superiority of the federal government over the states was critical to the idea of the American nation.

It was impossible to preserve both peace and the Union. The determina-tion of the federal government, and its position with regard to the states, was demonstrated when Lincoln refused to yield to demands for the surrender of the federal position of Fort Sumter in Charleston Harbor—the reality of national power in the face of the forge of Southern consciousness and sepa-ratism. On April 12, 1861, Confederate forces opened fire on the fort. Far from intimidating him into yielding, as Southern leaders had hoped, this clash led Lincoln to act, going to war to maintain the Union and not for the emancipation of the slaves. Lincoln's clear intention to resist secession with force by invading the Lower South played the major role in leading the Upper South, notably Virginia, to join the Confederacy, as it did not intend to provide troops to put down what Lincoln termed an insurrection.

In the seceding states, the prevalence of slavery varied greatly and this was linked to the degree of support for the war, although it was not the sole fac-tor involved. Many who fought for the South did not own slaves and were more motivated by a sense of the need to defend communities, cultures, and the states' rights that were believed to protect both. These states' rights were defined in part in terms of the defense of slavery. Slaves were cowed in a wave of terror, including lynchings. However, four slave states—Delaware, Maryland, Kentucky, and Missouri—stayed with the Union, as did those parts of Virginia that, in 1863, became the state of West Virginia.

The course of the war played a key role in the demise of slavery. Initially, the Union had anticipated a speedy victory and had not sought the radical change that Abolition entailed. In November 1862, the Union's envoy in Paris told the French Foreign Minister that "neither principle nor policy will induce the United States to encourage a 'servile war' or prompt the slave to cut the throat of his master or his master's family,"[81] a clear reference to deep-seated racial anxieties. Nevertheless, war goals and military methods changed in response to the difficulty and unexpected length of the conflict. Attacks by Union forces on private property in order to weaken the South

economically came to the fore from 1862. In the North, there were significant pressures on the home front, where frustration with the intractability of
the struggle led to an abandonment of conciliation toward Southerners in
late 1862. Initially, the Union had made no attempt to abolish slavery. This
was because Lincoln feared the impact of emancipation on sections of
Northern opinion, especially in loyal border states such as Kentucky.
Furthermore, like many others, he hoped that avoiding a pledge to support
emancipation would weaken Southern backing for secession.

After the Southern advance and run of success had been stopped at the
battle of Antietam on September 17, 1862, Lincoln, in contrast, heeded
radical Republicans, many of whom were linked to the Congressional Joint
Committee on the Conduct of the War, and the Union became committed
to the emancipation of the slaves in those parts of the South still in rebellion. Emancipation was regarded as a way to weaken the Southern economy, and thus war effort, as well as providing a clear purpose to maintain
Northern morale and a means to assuage the sin that was leading a wrath-
ful God to punish America. As with Britain against Napoleon in 1807,
Providence was to be recruited. The international audience was also in
Lincoln's mind. In 1862, the government agreed to a search treaty that
would hit the illegal slave trade under the American flag, a measure long
called for by Britain.

At the same time, many Republicans were against full equality, saw emancipation only as a wartime expedient, and wanted a postwar restriction of
black people essentially to the South or their dispatch abroad. The opposition
Democrats were against the end of slavery.[82]

Emancipation, like conscription, another radical step, was also linked to
the need for troops. Lincoln adapted his goal of preserving the Union to the
circumstances of the war, including the flight of slaves from their masters.
The recruitment of black people for the Union army was a symbol to, and
for, the Confederacy of what was a total war, and in the Union of a new citizenship.[83] Moreover, the recruitment of all-black regiments, numbering
more than 120,000 men, was also a major operational help, although black
soldiers were paid less well.

In the South, the slave basis of society collapsed as Union forces advanced
in 1864, with slaves seizing the opportunity to escape to servitude. As a
result, the black world was in considerable flux before the end of the war.
Union victory settled the matter. In 1865, the Thirteenth Amendment to the
Constitution ended slavery in the loyal as well as the Confederate states,
freeing about 4 million slaves. In 1866, Congress passed the Civil Rights Act,
giving full citizenship to all born or naturalized in the United States, and
voting rights to all male citizens. Black people thus gained legal equality.
The provisions of this legislation became the Fourteenth Amendment, which
was ratified in 1868.

Slavery Ends in Brazil

Prior to emancipation in 1888, the majority of black people in Brazil were already free, in part because of increased manumission under the Law of the Free Womb or Rio Branco law of 1871, which stated that all future children born to slave mothers would become free from the age of 21 (a clause that led to false registrations by owners), while in addition slaves were allowed to purchase their freedom. Slavery was regarded in influential circles, especially in the expanding cities, as a cause of unrest (which indeed increased in the 1880s), and a source of national embarrassment and relative backwardness. As part of the process by which New World settler societies were culturally dependent on the Old World, the elite looked to Europe to validate their sense of progress, and were affected by the extent to which slavery was increasingly presented as an uncivilized characteristic of barbaric societies and as incompatible with civil liberty. The abolition of slavery in the USA was part of this process.

Furthermore, slavery was seen by some, particularly the growing industrial lobbies, as an inefficient system compared to wage labor. The combination of the end of the slave trade with economic expansion meant that slavery was no longer able to supply Brazil's labor needs, including those in the traditional center of slavery, the North-East, and this situation helped make slavery seem anachronistic. Not only was quantity of labor an issue, but also type, as a growing need for artisans was not one that could be met from the traditional Brazilian slave economy, not least because that economy did not place much of an emphasis on educating slaves. European labor appeared more skilled, but, as an indication of the role of ideological factors including racism, this European labor also offered an additional degree of white admixture for the population mix, one that matched the changing priorities of the Brazilian elite which saw the whitening of the population as a source of stability.

The demand for more skilled labor in Brazil was an aspect of the degree to which modernization led to the demise of slavery, not only culturally and ideologically, but also for economic reasons. However, at the same time, Western economic growth had helped provide the demand, finance, and technological innovation that kept slavery a major option, and this is a reminder of the ambivalent relationship between modernization and slavery. Nevertheless, as the role of slavery in the Brazilian economy and of slaves in the net capital of Brazilian financial assets declined, so it seemed anachronistic and a legacy issue. As a result, slave owners became increasingly isolated, with free labor becoming more important even in some plantation areas, such as São Paulo, although there was also a continued preference for slave labor in others. The end of the slave trade had led to higher slave prices and a concentration of ownership,[84] and this reduced political support for slavery, a similar process to that in the American South. The higher prices and change in support lessened the stake of the wider economy in the slave trade.

In 1884, two Brazilian provinces emancipated slaves, creating free labor zones, and in 1885 all slaves over 60 were freed. Furthermore, increased numbers of slaves fled, many to the cities, such as Rio de Janeiro, so that by 1887 there were fewer than one million slaves—only about 5 percent of the Brazilian population. There was far more support for slave flight than in the USA, with much of the populace, as well as the bulk of the authorities, unwilling to support the owners. This situation contrasted markedly with that in the USA in 1861, a contrast that was very important to the subsequent history of the two countries. Because in Brazil the slave owners were without a mass domestic constituency, their eventual position was more similar to that of counterparts in the Caribbean than those in the American South.

In Brazil, the military was not keen on hunting escaped slaves, while, conversely, unlike in the USA, there was no serious regional separatism based on slavery. This was a key aspect of the largely nonviolent nature of Brazilian Abolitionism. There was no equivalent to the situation in Cuba, where the Ten Years' War of 1868–78, an unsuccessful independence struggle against Spanish rule, had seen partial abolition in rebel areas, which encouraged the move for gradual abolition in the island as a whole.[85]

Whereas in the American South there was a stress on white society as the people, a stress that, throughout the century, was to encourage racial exclusion, whether slave-based or not, as a form and focus of Southern cultural identity, in Brazil the emphasis, in contrast, was on a multicultural society, however much it was dominated by the white elite. In Brazil, the 1888 Golden Law, passed by an overwhelming majority in Parliament, freed the remaining slaves, about three-quarters of a million in total, without compensation. This law helped legalize the situation caused by large-scale flight, and has also been seen as an attempt to retain workers on the land.[86] In a related and important rejection of the past, Brazil became a republic the following year.

Conclusion

Brazil was the last of the slave societies in the Americas and the largest. That it abolished slavery only just over a century and a quarter ago indicates the significance of slavery for the Americas and the Atlantic world, and helps to explain why its legacy remains so pressing. In Brazil, slavery had been an integral part of society, whereas in the European empires there was a detachable quality and, in part as a product of this imperial distance from the colonies, it proved easier to secure political support for Abolition. The end of slavery in Brazil in part reflected developments that were of growing significance in the West, notably industrialization and democratization. Neither involved economic equality, but each led to a rethinking of the nature of citizenship. In this context, slavery appeared completely redundant. It disappeared with far less difficulty in Brazil than the case in the USA.

The case of Brazil in abolishing slavery in 1888 was very different from that of Britain and the slave trade in 1807. What, however, these cases show is that the end of the slave trade and of slavery reflected the range of factors that were in play in nineteenth-century culture, economics, and politics. Looked at differently, this range reflected the widespread resonance of both the slavery and the slave trade, indeed their significance, both for many aspects of life, but also for the way in which the world was considered. Many of these issues were to prove relevant for subsequent systems and practices of coerced labor.

Notes

1 J.R. Oldfield, *Transatlantic Abolitionism in the Age of Revolution: An International History of Anti-Slavery c.1787–1820* (Cambridge, 2013).
2 E. Gøbel, "The Danish Edict of 16th March 1792 to Abolish the Slave Trade," in *Orbis et orbem: Liber amicorum Jan Everaert*, eds. J. Parmentier and S. Spanoghe (Ghent, 2001), 251–63. More generally, see D. Eltis and J. Walvin, eds., *The Abolition of the Atlantic Slave Trade: Origins and Effects in Europe, Africa, and the Americas* (Madison, WI, 1981).
3 A. Sens, "Dutch Anti-Slavery Attitudes in a Decline-Ridden Society, 1750–1815," in *Fifty Years Later: Anti-Slavery, Capitalism and Modernity in the Dutch Orbit*, ed. G. Oostindie (Leiden, 1995).
4 D. Hempton, "Popular Evangelicalism and the Shaping of British Moral Sensibilities, 1770–1840," in *British Abolitionism and the Question of Moral Progress in History*, ed. D.A. Yerxa (Columbia, SC, 2012), 59–60.
5 C. Norris, "John Wesley and enslavement revisited," *Proceedings of the Wesley Historical Society* (2022) and report to the Methodist Church of Great Britain's Justice, Dignity and Solidarity Committee, Reparations Group, unpublished 2023. I would like to thank Clive Norris for sending a copy.
6 P.J. Marshall, *The Making and Unmaking of Empires: Britain, India, and America c. 1750–1783* (Oxford, 2005), 195; C. Kidd, *The Forging of Races: Race and Scripture in the Protestant Atlantic World, 1600–2000* (Cambridge, 2006).
7 M. Postlethwayt, *Universal Dictionary* (4th ed., 2 vols., London, 1774), I, no pagination, entry for Africa.
8 S. Wise, *Though the Heavens May Fall: The Landmark Trial that Led to the End of Human Slavery* (London, 2006); C.L. Brown, *Moral Capital: Foundation of British Abolitionism* (Chapel Hill, NC, 2006); N.S. Poser, *Lord Mansfield: Justice in the Age of Reason* (Montreal, 2013), 292–9.
9 R. Hanley and K. Gleadle, "The children's war on slavery," *BBC History Magazine* (July 2023), 53–7.
10 C. Midgley, *Women Against Slavery: The British Campaigns, 1780–1870* (London, 1992).
11 T. Odumosu, *Black Jokes, White Humour: Africans in English Caricature, 1769–1819* (London, 2017).
12 T.C. Hansard, *The Parliamentary History of England from the Earliest Period to the Year 1803*, vol. 29 (London, 1817), 278.

13 W. Woodfall, *Debate on a Motion for the Abolition of the Slave Trade in the House of Commons on Monday the Second of April 1792* (London, 1792), 41–2.

14 Ibid., 139.

15 Library of Congress, Washington, British caricature collection, 2–575.

16 *Archives parlementaires de 1790 à 1860: Recueil complet des débats législatifs et politiques des chambers françaises* (127 vols., Paris, 1879–1913), vol. 37, 152.

17 L. Clay, "Liberty, Equality, Slavery. Debating the Slave Trade in Revolutionary France," *American Historical Review*, 128 (2023), 117.

18 R.J. Alderson, *This Bright Era of Happy Revolutions: French Consul Michel-Ange-Bernard Mangourit and International Republicanism in Charleston, 1792–1794* (Charleston, SC, 2008); J. Popkin, *You Are All Free: The Haitian Revolution and the Abolition of Slavery* (New York, 2010).

19 W. Doyle, *France and the Age of Revolution* (London, 2013), 205.

20 S. Hazareesingh, *Black Spartacus: The Epic Life of Toussaint Louverture* (New York, 2020).

21 L. Dubois, *Avengers of the New World: The Story of the Haitian Revolution* (Cambridge, MA, 2004).

22 R.N. Buckley, *Slaves in Red Coats: The British West India Regiments, 1795–1815* (New Haven, CT, 1979).

23 E.L. Pierce, ed., *Memoirs and Letters of Charles Sumner* (London, 1878), 68–9.

24 J. Scott, *The Common Wind: Afro-American Current sin the Age of the Haitian Revolution* (London, 2018).

25 K.R. Maxwell, *Conflicts and Conspiracies: Brazil and Portugal, 1750–1808* (New York, 1973); G. Paquette, *Imperial Portugal in the Age of Atlantic Revolutions: The Luso-Brazilian World, c. 1770–1850* (New York, 2013), 104.

26 D. Turley, *The Culture of English Antislavery, 1780–1860* (London, 1991).

27 E. Williams, *Capitalism and Slavery* (Chapel Hill, NC, 1944). An influential work, particularly thanks to its 1960s' reissues in 1961, 1964, and 1966.

28 S.H.H. Carrington, *The Sugar Industry and the Abolition of the Slave Trade, 1775–1810* (Gainesville, FL, 2002).

29 S. Drescher, *Econocide: British Slavery in the Era of Abolition* (London, 1977).

30 R. Anstey, "Capitalism and Slavery: A Critique," *EcHR*, 2nd ser., 21, no. 2 (1968), 320.

31 K. Jacoby, "Slaves by Nature? Domestic Animals and Human Slaves," *Slavery and Abolition*, 15 (1994), 96–7.

32 S. Drescher, "Whose Abolition? Popular Pressure and the Ending of the British Slave Trade," *Past and Present*, 143 (1994), 136–66, esp. 165–6.

33 J. Coffey, "'Tremble Britannia!': Fear, Providence and the Abolition of the Slave Trade, 1758–1807," *English Historical Review*, 127 (2012), 844–81.

34 *House of Lords Debates*, February 5, 1807, vol. 8, columns 657–8, 662.

35 B.E. Vick, *The Congress of Vienna. Power and Politics after Napoleon* (Cambridge, MA, 2014), pp. 202–12.

36 A. Burton, "British Evangelicals, Economic Warfare and the Abolition of the Atlantic Slave Trade, 1794–1810," *Anglican and Episcopal History*, 65 (1996), 197–225, esp. 223.

37 C.J. Bartlett and G.A. Smith, "A 'Species of Milito-Nautico-Guerrilla-Plundering Warfare.' Admiral Alexander Cochrane's Naval Campaign against the United States, 1814–1815," in *Britain and America Go to War: The Impact of War and Warfare in Anglo-America, 1754–181*, eds. J. Flavell and S. Conway (Gainesville, FL, 2004), 187–90; John Harriott to Sidmouth, May 7, 1814, Exeter, Devon

Record Office, Sidmouth papers, 152M/C1814/OF13. See also 152M/C1813/OF3 and NA. War Office papers 1/141, 63–7.

38 G.A. Smith, *The Slaves' Gamble: Choosing Sides in the War of 1812* (New York, 2013).

39 D. Eltis, "The British Trans-Atlantic Slave Trade after 1807," *Journal of Maritime History*, 4 (1974), 1–11, and "The British Contribution to the Nineteenth-Century Transatlantic Slave Trade," *EcHR*, 2nd ser., 32 (1979), 211–27.

40 A. Chomsky, B. Carr, and P.M. Smorkaloff, eds., *The Cuba Reader* (Durham, NC, 2003), 37.

41 Cf. 50–5, drawing on E.P. Garfield (trans.), *Autobiography of a Slave* by Juan Francisco Manzano (Detroit, MI, 1996).

42 H. Klein and S. Engerman, "Shipping Patterns and Mortality in the African Slave Trade to Rio de Janeiro, 1825–1830," *Cahiers d'études africaines*, 15 (1975), 385–7.

43 J.P. Marques, *The Sounds of Silence: Nineteenth-Century Portugal and the Abolition of the Slave Trade* (Oxford, 2006).

44 R.J. Reid, *Political Power in Pre-Colonial Buganda: Economy, Society and Warfare in the Nineteenth-Century* (London, 2002).

45 K.O. Dike, *Trade and Politics in the Niger Delta 1830–1885* (Oxford, 1956), 68–9.

46 R. Dunn, *A Tale of Two Plantations: Slave Life and Labor in Jamaica and Virginia* (Cambridge, Mass., 2014).

47 J.C. Dorsey, *Slave Traffic in the Age of Abolition: Puerto Rico, West Africa, and the Non-Hispanic Caribbean, 1815–1859* (Gainesville, FL, 2003).

48 J.D. Martin, *Divided Mastery: Slave Hiring in the American South* (Cambridge, MA, 2004).

49 J. Powell, *Losing the Thread: Cotton, Liverpool and the American Civil War* (Liverpool, 2021).

50 C. Tomlins, *In the Matter of Nat Turner: A Speculative History* (Princeton, NJ, 2020).

51 M. Echeverri, *Indian and Slave Royalists in the Age of Revolution: Reform, Revolution and Royalism in the Northern Andes, 1780–1825* (Cambridge, 2016).

52 R.J. Cottrol, *The Long, Lingering Shadow: Slavery, Race, and Law in the American Hemisphere* (Athens, GA, 2013).

53 R. Conrad, *World of Sorrow: The African Slave Trade to Brazil* (Baton Rouge, LA, 1986).

54 L. Bethell, *The Abolition of the Brazilian Slave Trade* (New York, 1970).

55 D.R. Murray, *Odious Commerce: Britain, Spain and the Abolition of the Cuban Slave Trade* (Cambridge, 1980).

56 P.M. Kielstra, *The Politics of Slave Trade Suppression in Britain and France, 1814–48: Diplomacy, Morality and Economics* (Basingstoke, 2000).

57 P. Grindal, *Opposing the Slavers: The Royal Navy's Campaign against the Atlantic Slave Trade* (London, 2016).

58 D.L. Canney, *Africa Squadron: The U.S. Navy and the Slave Trade, 1842–1861* (Washington, DC, 2006); C.H. Gilliland, ed., *U.S.S. Constellation on the Dismal Coast: Willie Leonard's Journal, 1859–1861* (Columbia, SC, 2013).

59 A.A. Boahen, *Britain, the Sahara and the Western Sudan, 1788–1861* (Oxford, 1964).

60 Stevenson to Liverpool, February 1, 1812, Exeter, Devon County Record Office, 152H/C1812/OF27.

61 N. Thompson, *Earl Bathurst and the British Empire* (Barnsley, 1999), 167–8.

62 J.J. Ewald, *Soldiers, Traders and Slaves: State Formation and Economic Transformation in the Greater Nile Valley, 1700–1885* (Madison, WI, 1990).

63 M. Crowder and S. Miers, "The Politics of Slavery in Bechuanaland: Power Struggles and the Plight of the Basarwa ..." in *The End of Slavery in Africa*, eds. S. Miers and R. Roberts (Madison, WI, 1988), 175–6.

64 J. Lean and T. Burnard, "Hearing Slave Voices: The Fiscal's Reports of Berbice and Demerara-Essequebo," *Archives*, 27 (2002), 122.

65 A.J. Barker, *Slavery and Antislavery in Mauritius, 1810–33: The Conflict Between Economic Expansion and Humanitarian Reform under British Rule* (London, 1996); B.W. Higman, *Slave Populations of the British Caribbean, 1807–1834* (Baltimore, MD, 1984).

66 J.R. Ward, *British West Indian Slavery, 1750–1834: The Process of Amelioration* (Oxford, 1988).

67 P. Dumas, *Proslavery Britain: Fighting for Slavery in an Era of Abolition* (Basingstoke, 2016).

68 M. Turner, *Slaves and Missionaries: The Disintegration of Jamaican Slave Society, 1787–1834* (Urbana, IL, 1982); E.V. d'Costa, *Crowns of Glory, Tears of Blood: The Demerara Slave Rebellion of 1823* (New York, 1994).

69 M. Barcia, *West African Warfare in Bahia and Cuba: Soldier Slaves in the Atlantic World, 1807-1844* (Oxford, 2014).

70 J.D. Milligan, "Slave Rebelliousness and the Florida Maroon," *Prologue*, 6 (spring 1974), 4–18.

71 M.C. Karasch, *Slave Life in Rio de Janeiro, 1808–1850* (Princeton, NJ, 1987); B.J. Barickman, *A Bahian Counterpoint: Sugar, Tobacco, Cassava, and Slavery in the Recôncavo, 1780–1860* (Stanford, CA, 1998).

72 I. Gross, "The Abolition of Negro Slavery and British Parliamentary Politics, 1832–3," *Historical Journal*, 23 (1980), 84.

73 C. Hall, N. Draper, K. McClelland, K. Donington and R. Lang, *Legacies of British Slave-Ownership: Colonial Slavery and the Formation of Victorian Britain* (Cambridge, 2014).

74 R. Huzzey, *Freedom Burning: Anti-Slavery and Empire in Victorian Britain* (Ithaca, NY, 2012).

75 S. Drescher, *From Slavery to Freedom: Comparative Studies in the Rise and Fall of Atlantic Slavery* (New York, 1999).

76 L. Dubois, "The Road to 1848: Interpreting French Anti-Slavery," *Slavery and Abolition*, 22 (2001), 150–7.

77 L. Brophy, *University, Court, and Slave: Pro-Slavery Thought in Southern Colleges and Courts and the Coming of the Civil War* (New York, 2016); E. Herschthal, *The Science of Abolition: How Slaveholders Became the Enemies of Progress* (New Haven, CT, 2021).

78 R. Fogel, *Without Consent or Contract: The Rise and Fall of American Slavery* (New York, 1990).

79 Lyons to Lord John Russell, the Foreign Secretary, December 12, 1859, NA. PRO. 30/22/34 fol. 69.

80 C.L. Paulus, *The Slaveholding Crisis: Fear of Insurrection and the Coming of the Civil War* (Baton Rouge, La., 2017).

81 NA. FO. 5/877 fol. 149.

82 P. Escott, *The Worst Passions of Human Nature: White Supremacy in the Civil War North* (Charlottesville, VA, 2020).

83 D. Willis, *The Black Civil War Soldier: A Visual History of Conflict and Citizenship* (New York, 2021).

84 Drescher, "Brazilian Abolition in Comparative Perspective," *Hispanic American Historical Review*, 68 (1988), 433.

85 A.F. Corwin, *Spain and the Abolition of Slavery in Cuba, 1817–1886* (Austin, TX, 1967).

86 R.B. Toplin, *The Abolition of Slavery in Brazil* (New York, 1972); R. Conrad, *The Destruction of Brazilian Slavery, 1850–1888* (Berkeley, CA, 1972).

6   After Slavery?

The official end of the slave trade and slavery in the Western world was an aspect of the modernization, indeed modernity of this world, in the nineteenth century. At the same time, modernization was a more complex process than is suggested by that remark. The end of slavery did not completely transform labor relations or indeed racial politics in their broadest sense, whether in Britain's Caribbean colonies, or in other former slave societies such as Haiti and Brazil.[1] Control over labor continued. This control was seen from the outset. In Haiti, the plantation economy producing cash crop for European markets survived black independence in 1804. Slavery had gone, but the black elite who ran the state used forced labor to protect their plantations from the preference of people to live as peasant proprietors. Thus, in Haiti, the pressures of the global economy, and the attractions of cash crops selling into international markets, not least in order to meet French pressure for compensation, triumphed over the potential consequences of independence.[2]

The Situation in the Former Colonies

Control, formal or informal, continued to play a key role whatever the formal status of workers. In the British colonies, as elsewhere, many former slaves were pressed into continuing to work in sugar production.[3] Legal systems were employed in order to limit the mobility and freedom of former slaves—for example, by restricting emigration and also what was presented as vagrancy. Rents were also used to control labor and to reduce labor costs. Resistance to this labor control included strikes as well as leaving the plantations, while, in contrast, the new system was supported by punishment, including the use of workhouses where the harsh regime acted as a potent form of control. For example, once female apprentices entered the workhouse, they were no longer protected by the Abolition Act.[4] The difficult situation for workers in the British colonies after 1838 undercuts any simple attempt to create a contrast between slavery and freedom.[5] Indeed, the

DOI: 10.4324/9781003457923-6

conditions of labor for slaves and ex-slaves reflected far more than the legal situation.[6] Across the world, for most former slaves, there was no sweeping change in their lives as a result of the abolition of slavery. Instead, many former slaves remained dependent, in some way or other, on their ex-masters or on new masters. The very flexibility of economic service and subjugation ensured the continuation of systems of labor control, and these encompassed labor flows. The same was true of Russia, where in 1861 Tsar Alexander II emancipated the serfs.

Moreover, despite the abolition of the slave trade and slavery, labor continued to flow to the colonies, both British and other. This was because, despite attempts to control them, former slaves tended to take up small-scale independent farming on provision grounds, rather than work on plantations, and this situation helped lead to demands for fresh labor. In place of slaves, the British West Indies, especially Trinidad, British Guiana, and other colonies, received cheap Indian indentured labor, although, despite its availability, sugar production declined.[7] For example, nearly a quarter of a million indentured Indians moved to British Guiana (now Guyana) from 1838 to 1918, and 150,000 to Trinidad from 1845 to 1917. Critics claimed that the indentured labor systems, which were also employed in Cuba and the French Caribbean, represented the continuation of the slave trade in its latter stages, not least due to the coercive character of these systems.

At the same time, West Indians found it necessary to travel abroad for work. West Indians provided much of the labor for attempts to dig canals across the isthmus of Panama. In the 1880s, the French project relied largely on Jamaican labor, while the American direction that successfully brought the new canal project to completion in 1914 relied heavily on labor from Barbados. About 45,000 out of Barbados's population went to Panama where they were harshly treated and had a far higher death rate than other workers.

Dependence on new or former masters was also the case for many former slaves in the USA. A new order had certainly been seen with Union victory in 1865. The recruitment of all-black regiments for the Union army in the second half of the Civil War, numbering more than 120,000 men, was both a major operational help to the Union, and also a symbol to what to the Confederacy was indeed a total war. Although there was a widespread lack of white interest in their achievement, black troops were given combat roles, the action at Fort Wagner in July 1863 proving a key watershed, and could be the majority of a force. The symbolic power of black troops was shown in 1865 when the forces that occupied Charleston, the site of the outbreak of the war, included black troops recruited from former Carolina slaves.[8] Larger numbers of slaves had escaped as the advance of Union forces brought disruption to the South, not least with General Sherman's advance across Georgia in 1864.

Union victory led to a new order in the South, reflecting the overthrow, at least in law, of the previous system of exclusion, subordination, and oppression. In 1865, the Thirteenth Amendment to the Constitution led to the freeing of about 4 million slaves. In 1866, over the veto of President Andrew Johnson, who had been a slave owner and who sought reconciliation with white Southerners, Congress passed the Civil Rights Act, giving full citizenship to all born or naturalized in the USA, and voting rights to all male citizens. Black people thus gained legal equality. The provisions of this legislation became the Fourteenth Amendment, which was ratified in 1868. These amendments reflected the extent to which the radicalizing nature of the war, and the issues to which it had given rise, made it possible to envisage improvements on the provisions decreed by the Founding Fathers.

However, in the face of strong and often violent pressure from white Southerners, the Reconstruction Acts of 1867, which dissolved the Southern state governments and reintroduced federal control, were not sustained. The foundation of the Ku Klux Klan, a Confederate veterans' movement, in 1866 was followed by several thousand lynchings, although not all were by the Klan. Federal troops were withdrawn from the South in 1877, the black militias recruited by Radical Republican state governments lost control or were disbanded, and the black people in the South were very much left as second-class citizens, vulnerable to economic and political exploitation.[9]

Exacerbated by the rise of *de jure* segregation in the South in the 1890s, this situation persisted until the 1950s and 1960s. The South became a poor agricultural society with oppressive labor relations.

In its 1896 *Plessy v. Ferguson* judgment, the Supreme Court argued that:

The object of the statute [14th Amendment] was undoubtedly to enforce the absolute equality of the two races before the law, but in the nature of things it could not have been intended to abolish distinctions based upon color, or to enforce social, as distinguished from political equality, or a commingling of the two races upon terms unsatisfactory to either.

Meanwhile, despite the introduction of Indian indentured labor, the former plantation societies of the West Indies and British Guiana became far less important to the British economy. This was a process accentuated by the equalization of the sugar duties in 1846, under the Sugar Duties Act. The act was an aspect of a more general rejection of protectionism as part of the adoption of Free Trade as both economic policy and fiscal policy. Under the act, protection for British sugar was progressively reduced until all duties on imported sugar were equalized in 1851.[10] In practice, the act encouraged sugar imports into Britain from Brazil and Cuba, which hit the British sugar producers. The abolition of slavery, therefore, played a major role in a more general crisis of the British plantation societies, a crisis which was particularly marked in Jamaica. Labor availability and discipline were crucial in the ability of plantations to hold down costs. With Emancipation, productivity and profitability fell, as, despite the major possibilities offered by the adoption of steam power in the shape of steam milling, sugar production continued to be labor-intensive. Free labor proved more expensive and less reliable than slaves, greatly increasing the operating costs. In Jamaica, 314 estates, 49 percent of the sugar plantations, ceased cultivation between 1844 and 1854, and there was insufficient investment in the others. Similarly, in Brazil, the end of slavery in 1888 hit the sugar economy of the north-east as a key aspect of a more general agrarian depression that affected the old order in Brazil, weakening the imperial monarchy. Brazil became a republic in 1889.

As the exports of the former British plantation economies declined, so they were less able to attract investment, afford imports from Britain and elsewhere, and develop social capital. This had a major impact on the living standards of the bulk of the population of these colonies, one that should be considered when the present-day situation in the West Indies is debated. In 1815, the West Indies had been the leading market for British exports, but by 1840 it had been passed as a market by India, Australia, and Canada, in that order.[11] Linked to this, the role of the West Indies in British shipping needs also diminished. The decline of the plantation economies indeed helped ensure that the share of the empire in British trade fell, although the expansion of British trade with other countries outside the empire was also important.

Moreover, in the former slave colonies, the problems, including racism centered on slavery, had changed, not ended. This was to be made clear in Jamaica by the harsh (and illegal) suppression of the Morant Bay "uprising" in 1865. In this episode, resentment among poor, black peasantry led several hundred protestors under Paul Bogle, a black deacon, to seek to petition the Governor, Edward Eyre, only for the militia to respond violently, killing seven protestors. In the resulting riots, 18 planters were killed. Fearing an uprising of the 350,000 black people against 13,000 white people, Eyre imposed martial law for a month. Bogle and over 400 rebels were hanged and 1,000 homes burned. George Gordon, a black member of the Assembly critical of Eyre, was court-martialed and executed.

Impact on Africa

The end of the transatlantic slave trade also led to the development of plantation economies in parts of Atlantic Africa, particularly the large Portuguese colony of Angola. This development represented a response to labor availability in Africa. It also reflected a shift in the terms of Western trade with Africa, away from a willingness to pay for labor in the shape of slaves, and toward a willingness to pay for it in the form of products, and thus of labor located in Africa. There continued indeed to be multiple overlaps between servitude and trade in the Atlantic African economy. As one aspect, the slave trade was followed, although not at once, by colonialism. This colonialism was based on the exports of "legitimate commerce" from the areas of Africa now seized as colonies by European powers.[12] Such a transformation was an aspect of the extent to which the globalization of the late nineteenth century integrated much of the world into a system, of capital and trade, that reproduced cycles of dependence.[13] Racism was part of the equation as it helped justify imperial rule and colonial control, rather as it had helped justify the slave trade. As most of Africa was brought under at least formal European imperial control in the 1880s and 1890s, the Europeans were best placed to control the subsequent commercial economy.

In West Africa, the earlier end of the slave trade had hit those kingdoms and colonial bases that had derived wealth from it. However, the slave economy remained significant. In addition, it had considerable dynamism. For example, Kano, the metropolis of the Sokoto caliphate in what was to become northern Nigeria, had a population that, by the 1890s, had grown to about 100,000, half of them slaves. It was a major center of agriculture and industry, with leather goods and textiles manufactured and exported to much of the *sahel* as well as to North Africa. This activity drew on slave labor, including, in a practice longstanding in Islamic societies, for the Sokoto army that, itself, produced more slaves through raiding. A similar pattern could be seen with other industrial cities in the *sahel*.[14] The significance of slave soldiers was also seen with the impressive Egyptian army in the first half of the century, the slaves obtained by raiding in Sudan.

The slave trade increased in and from East Africa for most of the nineteenth century, seriously affecting much of the interior of the continent as it fed the coast.[15] This increase in part reflected the slave-based plantation systems that developed on Africa's Indian Ocean coast, producing, for example, cloves, the export of which boomed in the 1810s–40s. Furthermore, the export of slaves to the Islamic world—for example, Arabia—continued. Kilwa was the leading port in East Africa, with a major trade from there to the Gulf, much of it handled by Omani merchants who had strong links with the east coast, notably Zanzibar. Their profits helped finance economic activity, which, in turn, produced needs for slave labor. Slaves to Arabia, in contrast, were moved relatively short distances, particularly from the Red Sea ports of Suakin (modern Sudan) and Massawa (modern Eritrea), which were fed by slave trading and raiding into Sudan and Ethiopia.

Slave labor in East Africa was largely, as in the New World, plantation labor, but the situation was different in Arabia and the Gulf. This serves as a reminder of the economics of slavery and the variety of slave conditions, and therefore the slave experience. Slaves in Arabia and the Gulf were used for many activities, but, although slaves were used for date plantations in

Oman, there was no dominance by large-scale institutions of the plantation type. Instead, slaves were a mobile-labor force that was appropriate to different needs. The presence of slaves and of those of slave descent posed issues of identity in terms of the character of the people and how it was described legally and in other sources.[16]

As slaving was largely brought to an end in the Atlantic in the 1860s, so the British struggle against it in East African waters became more prominent, with the Cape Squadron being allocated to the task and, in 1864, merged with the East Indies Squadron. This struggle with Arab slavers was presented in an heroic light and was fully covered in British publications, including newspapers. It was also considered worth publishing books about the struggle, as in Captain George Sulivan's *Dhow Chasing in Zanzibar Waters* (1873) and Captain Philip Colomb's *Slave Catching in the Indian Ocean* (1873).

Opposition to slavery was not restricted to the oceans, but also affected British policy in Africa. The slaves of Indian traders in East Africa were confiscated by the British Consul in 1860 as the traders were British subjects. Moral activism towards Africa, especially hostility to the slave trade in Central and East Africa, had a number of consequences. These included the development of a British presence in Zanzibar,[17] and a strengthening of the determination to blaze the trail for Christian grace and morality. This determination was seen in the actions of, and response to, David Livingstone, a prominent missionary who helped secure British pressure to persuade the Sultan of Zanzibar to outlaw the slave trade in 1873. HMS *London*, an old ship of the line, was sent to Zanzibar to enforce this prohibition, and the acquisition from Germany of Zanzibar itself as a British protectorate in 1890, in exchange for the island of Heligoland in the North Sea, led to the end of the trade.

This is an optimistic account of the impact of British anti-slavery in Africa, as it is possible to emphasize the extent to which hostility to slavery helped inspire, direct, and provide an explanation for British concern, interests, and British imperialism in Africa. This was the case both at the expense of local interests and of those of other imperial powers. The role of anti-slavery in British imperialism thus ensured that it played a part in the rankings of peoples and justifications of control that inherently reflected and entailed racism and coercion. Moreover, there was a willingness at the same time on the part of Britain to develop indentured labor systems and other practices of labor control.[18] Racism, indeed, took on new intellectual energy as the idea of permanent, biologically based differences between the races was rephrased in terms of the ethnology, indeed "race science," that became more significant from the mid-nineteenth century. Monogenesis, with its basis in the idea of the common origin of all human varieties in Eden, was challenged by polygenism, the belief in the separate origin of different human "races."

States for Former Slaves

As a different form of activity and influence, colonies for free Africans, many freed slaves, had been founded. These colonies faced the difficulty that philanthropic goals were not generally seen in that light by local Africans who were suspicious of attempts to occupy their land.[19] Sierra Leone, established by the British in 1787, was the first. Both the Committee for the Relief of the Black Poor and key government supporters appear to have been motivated in founding this colony by humanitarianism springing from Christian convictions, gratitude towards black Loyalists from the former North American colonies, and Abolitionist sympathies; and the settlement explicitly forbade slavery. The great majority of newspaper items covering the expedition were sympathetic in tone. Combined with intermarriage and the good public response to the appeal for money to help poor blacks, this suggests that racial hostility may have been less common than has often been assumed.[20] Subsequently, slaves freed by the British navy were taken to Sierra Leone. The establishment of Sierra Leone indicates the range and energy of Abolitionist activity at this juncture, both in the 1780s and thereafter.

The British were not alone. In 1849, the French founded Libreville in Gabon for freed slaves. Inspired by Sierra Leone, the American equivalent, Liberia, was originally established by an anti-slavery group, the American Colonization Society (founded in 1816), as a home for freed American slaves. The first settlers landed in 1821. In part, Western backing for these states, notably in the USA, rested on a support for racial removal, a support that could align Abolitionist sympathies with a hostility to multiculturalism. Colonists were taken to Liberia by American warships. The American government gave scant support to Liberia, whereas Sierra Leone, which had become a British crown colony in 1808, saw the government back the Church Missionary Society in building schools and spreading Christianity.[21]

Initially a colony, Liberia declared independence in 1847. Once established, the free black Americo-Liberians harshly treated the local Africans. They used them as a source of labor to produce palm oil that was shipped to the USA. Palm oil and other trade goods were often produced or transported by slaves. Moreover, as earlier with Western slavers, there was a clash between the Americo-Liberians and the local Africans over the African practice of pawning. In this case, the Africans treated pawning as, for example, a form of apprenticeship designed to help children acquire skills, while the new settlers saw it as a form of controlled labor.

Abolishing Slavery

In Africa, the treatment of slaves held within native society became more problematic as the Westerners became colonial rulers across most of the continent. As with the Russians in Central Asia, however much other factors also played a role,[22] they were committed to Abolition as part of the civilizing mission used to justify imperialism as progressive. This approach was underlined by the Congress of Berlin in 1885 and by the anti-slavery conference at Brussels in 1889. This mission, however, risked offending local vested interests that were seen as crucial to the stability of imperial control—for example, that of the British in northern Nigeria where they established a protectorate in 1900. There were also concerns about the economic impact of Abolition.

Nevertheless, slavery was largely ended, or at least driven into relative obscurity, both by colonial policy and thanks to the slaves' attempts to seek advantage from their changing environment, a situation that paralleled that in the American South during the Civil War. The continuation of slave raiding was in 1902 a pretext for Britain going to war with the Sokoto caliphate in northern Nigeria. The following year, at Burmi, the Sokoto army was defeated, and the caliph and his two sons were killed. Resistance in northern Nigeria to British rule came to an end.[23]

The Twentieth Century

Slavery on the traditional pattern continued throughout the twentieth century in parts of Africa and Asia—for example, Mauritania in the former and Saudi Arabia in the latter. Some of these slaves were "trafficked" (i.e. traded, from a distance, notably from sub-Saharan Africa). War and debt proved prime sources of slaves. Thus, the warfare in southern Sudan from the 1960s to the 2010s led to the enslavement of captured men and women. This was encouraged by the role of ethnic hatred in conflicts there and elsewhere. Across Africa, rebel groups captured women for sex slaves. In 2014, Kashim Shettima, the Governor of Borno state in Nigeria, claimed that 230 schoolgirls, Christian and Muslim, recently kidnapped at gunpoint from their boarding house by the Islamic militant group Boco Haram, would be used as sex slaves and cooks. Subsequently, Abubakar Shekau, the head of the group, announced in a video "I abducted your girls. I will sell them in the market by Allah. There is a market for selling humans. Allah says I should sell. He commands me to sell. I will sell women. I sell women." Indeed, some were sold as brides, for £8 each, and smuggled to the neighboring countries of Chad and Cameroon. Adding a religious dimension by focusing on the Christian girls, Matthew Owojaiye, a spokesman for the Christian Association of Nigeria, commented: "Daughters of Zion taken captive, to be treated as slaves and sold into marriage to unclean people. Abomination has been committed." Boco Haram pressed on in 2014 to kidnap men for enforced service including as soldiers.

State Slavery

At the same time, there was a major increase in state slavery during the twentieth century, notably in the Soviet Union, the Nazi Empire, and North Korea. The first two cases involved major movements of people. For example, those sent to the *gulags*, or Soviet labor camps, were moved long distances. In 1939–40, about 1.4 million people were deported from Soviet-occupied Poland and from the occupied Baltic states to Soviet labor camps, many of which were in the Arctic. This was but one instance of a long-term movement based on the seizure of people as labor and for punishment. For example, the entire population of Crimean Tartars—an estimated 191,000—was deported to Soviet Central Asia on May 21, 1944 as punishment for alleged collaboration with the German occupiers of Crimea in 1942–4. They were loaded onto train wagons and kept there in crowded and diseased circumstances. Many died on the month-long journey. About 108,000 of the Crimean Tartars are estimated to have died of hunger, cold, and disease during their exile. Many were used as forced labor in the cottonfields and beaten by overseers when their quotas fell short.

Similarly, the Nazis moved large numbers of people to provide a workforce, mostly in Germany but also in German-controlled Europe. Aside from prisoners of war, many of whom were used as workers, 5.7 million foreign workers were registered in the Greater German Reich in August 1944. After brutally crushing the Warsaw Rising in 1944, the Germans in October 1944 deported about 150,000 of the city's Polish population to Germany in order to provide forced labor.

Holocaust

By 1944, most of the Jews from German-occupied Europe had been murdered. They earlier had provided slave labor. The movement of much of this labor was, as also in the Soviet case and that of the Japanese empire, extremely harsh, with scant effort made to keep people alive. This was "total labor," a form designed to lead to death. The use of Jewish labor indicates the difficulties of determining what slavery and the slave trade means in a more modern context than that of transatlantic trade flows. This use requires analysis because some of the popular modern discussion of the slave trade presents it as analogous to the Holocaust. In practice, this is a flawed comparison, as the Germans planned to kill all Jews, whereas the Atlantic slave-shippers and owners wished to keep slaves alive for labor and had no such genocidal intentions.

However, the theme of forced labor offers an instructive parallel. The forced labor of Jews became important in Germany from 1938 because Nazi persecution, in seizing Jewish businesses and closing down Jewish employment opportunities, made it possible to direct Jewish labor. Having made

Jews unemployed, they were given state welfare only on condition that they accepted employment in difficult and demeaning conditions that were designed to remove them from fellow Germans. Thus, Jews were made to work on processing rubbish or in projects in which they were segregated in camps. From 1939, this program expanded in response to the need by the army for non-Jewish German manpower. As a result, the significance of Jewish labor increased, and notably after the Germans conquered Poland that year. The ghettos, forced-labor camps, and concentration camps in which Polish Jews were confined were crucial elements in the network of compulsory work. For example, the ghetto in the city of Lodz, which was established in 1940, specialized in uniforms for the German army as well as other military supplies. Moreover, located in a key economic zone, that of coal-rich Upper Silesia, Auschwitz was not only an extermination camp, indeed the one where the largest number of Jews was gassed, but also a center of slave labor for the German war machine. The nearby large plant for the manufacture of synthetic rubber and oil was one of the largest German industrial projects. Across Germany and occupied Europe, the SS dominated the control of Jewish labor. Jews were made to work for the SS, other state agencies, and for private companies. For example, Organization Todt, the key construction agency for the German war machine, was responsible for the underground facilities built to safeguard weapons production from Allied air attack, as with the Nordhausen factory manufacturing V-weapons. The Jewish workers there were mistreated, many were killed, and others were sent to the extermination camp for gassing.

Jews moved to labor camps were treated in a particularly brutal fashion, and, to ensure that they could be, they were segregated in distinct teams. The food was limited and poor quality, the barracks were not heated, sanitary facilities were limited, clothing was inadequate, and the work was hard. Shifts were long; there was no concern with safety; many of the workers lacked relevant experience and there were frequent beatings and shootings. Jews were killed when ill, or worked to death. The intent was genocidal. Slave labor thus describes the very one-sided nature of control experienced by Jews. At the same time, the term does not suggest any equivalence with other slave systems and this point needs emphasizing in a context in which terms are frequently employed with inadequate precision.

Conclusions

That current-day slavery was to become more prominent anew in the 2000s and 2010s, not least as an issue, reflected not so much state slavery, although that was significant in North Korea, but, rather, labor flows in the world economy, as well as the hardships, opportunities, and exploitation that encouraged people to move, and to be moved and used. This harsh situation, however, has yet to produce an activism and, even more, an international and domestic governmental drive to match the cruelty involved.

Notes

 1 S. Miers, *Slavery in the Twentieth Century: The Evolution of a Global Problem* (Walnut Creek, CA, 2003).
 2 A. Dupuy, *Haiti in the World Economy: Class, Race, and Underdevelopment since 1700* (Boulder, CO, 1989).
 3 O.N. Bolland, "Systems of Domination after Slavery: The Control of Land and Labor in the British West Indies after 1838," *Comparative Studies in Society and History*, 23 (1981), 591–619, esp. 612–17.
 4 H. Altink, "Slavery by Another Name: Apprenticed Women in Jamaican Workhouses in the Period 1834–8," *Social History*, 26 (2001), 40–59, esp. 58.
 5 W. Kloosterboer, *Involuntary Labour Since the Abolition of Slavery* (Leiden, 1960); D. Eltis, ed., *Coerced and Free Migration: Global Perspectives* (Stanford, CA, 2002).
 6 P. Morgan, "Work and Culture: The Task System and the World of Lowcountry Blacks, 1700 to 1880," *William and Mary Quarterly*, 3rd ser., 39 (1982), 563–99.
 7 M. Kale, *Fragments of Empire: Capital, Slavery and Indian Indentured Labor Migration in the British Caribbean* (Philadelphia, PA, 1998).
 8 J.D. Smith, ed., *Black Soldiers in Blue: African American Troops in the Civil War Era* (Chapel Hill, NC, 2002).
 9 W. Rogers, *Reconstruction Politics in a Deep South State: Alabama, 1865–1874* (Tuscaloosa, AL, 2001).
10 W.A. Green, "The Planter Class and British West Indian Sugar Production Before and After Emancipation," *EcHR*, 2nd ser., 26 (1973), 448–63, esp. 462–3.
11 P.J. Cain, "Economics and Empire: The Metropolitan Context," in *The Oxford History of the British Empire, III: The Nineteenth Century*, ed. A. Porter (Oxford, 1999), 34–5.
12 R. Law, S. Schwarz, and S. Strickrodt, eds., *Commercial Agriculture, the Slave Trade and Slavery in Atlantic Africa* (Woodbridge, 2013).
13 R. Goddard, "In Praise of the Caribbean Plantation?" *Journal of the Historical Society*, 13 (2013), 462; D. Tomich and P.E. Lovejoy, eds, *The Atlantic and Africa: The Second Slavery and Beyond* (New York, 2021).
14 J. Osterhammel, *The Transformation of the World: A Global History of the Nineteenth Century* (Princeton, NJ, 2014), 293.
15 E. Alpers, *Ivory and Slaves in Central Africa* (London, 1975); A. Sheriff, *Slaves, Spices and Ivory in Zanzibar* (London, 1987); T. Ricks, "Slaves and Slave Traders in the Persian Gulf, 18th and 19th Centuries: An Assessment," in *The Economics of the Indian Ocean Slave Trade in the Nineteenth Century*, ed. W.G. Clarence-Smith (London, 1989), 60–70.
16 B.A. Mirzai, "African presence in Iran: Identity and its Reconstruction in the 19th and 20th Centuries," *Revue française d'outre-mer*, 89 (2002), 336–7.
17 R. Coupland, *The Exploitation of East Africa, 1856–90: The Slave Trade and the Scramble* (London, 1939).
18 R. Huzzey, *Freedom Burning: Anti-Slavery and Empire in Victorian Britain* (Ithaca, NY, 2012).
19 B.G. Smith, *Ship of Death: A Voyage that Changed the Atlantic World* (New Haven, CT, 2013), 80.

20 S.J. Braidwood, *Black Poor and White Philanthropists: London Blacks and the Foundation of the Sierra Leone Settlement, 1786–1791* (Liverpool, 1994).

21 B. Everill, *Abolition and Empire in Sierra Leone and Liberia* (New York, 2013).

22 J. Eden, *Slavery and Empire in Central Asia* (Cambridge, 2018).

23 R.H. Dusgate, *The Conquest of Northern Nigeria* (London, 1985); D. Williams, *I Freed Myself: African-American Self-Emancipation in the Civil War Era* (New York, 2014).

7 Conclusions

Slavery and the slave trade are contentious issues, and rightly so. Charges of exploitation and of historic wrongs explaining present circumstances are frequently advanced in a process that became stronger from 2020. These changes are important to the public memory in many countries, both in terms of their own history and also with reference to other states. This public memory is particularly significant because much of it relates to the related questions of the impact of the West and the nature of the two greatest global powers of the last three centuries, Britain and America. Moreover, there is no doubt that the slave trade played a formative role, not only in the demographics of the Atlantic world, but also of its varied political cultures and collective memories. These topics will be addressed in this chapter. There is also discussion of the extent to which slavery and the slave trade continue today.

Legacies

Liverpool, Britain's major slaving port when the slave trade was at its height in the eighteenth century, made a public apology in 1994 for its role in the slave trade and the International Slavery Museum in Liverpool is very candid about the city's involvement in slavery. A descendant of John Hawkins, England's most prominent slave trader in the sixteenth century, and a group of 20 friends, locked themselves in chains in 2006 in Gambia in West Africa in order to demonstrate their sense of sorrow, before being forgiven by the Vice-President. The Church of England apologized in 2006 to the descendants of victims of the slave trade, the General Synod acknowledging the "dehumanising and shameful" consequences of slavery. Indeed, in his 1957 sermon "The Birth of a New Nation," Martin Luther King Jr. had noted the role of the Church of England in giving slavery moral stature. In 2007, Tony Blair, the British Prime Minister, expressed "deep sorrow and regret" for the "unbearable suffering" caused by Britain's role in slavery, although stopping short of a formal apology.

DOI: 10.4324/9781003457923-7

International movements delight those who like to find commonalities in cause, course and consequence, but each country has a unique dimension in every crisis and there is danger to reading readily from one to another. And so also with Britain. The demonstrations, agitation and commentary linked to the Black Lives Matter controversy seen in 2020, notably in Bristol and London but in practice across much of the country, saw both deeper and more widespread tendencies and ones specific to Britain, particularly to the contentious legacy of empire.[1] The latter provided a matter of intellectual and conceptual confusion on the part of much of the agitation, with an elision of the distinction between discussion of the slave trade and that of the empire. In reality, the two were very different, and one of the major activities of the empire was the campaign against slavery. That distinction, however, was of no interest to what rapidly became a movement drawing together a range of interrelated discontents.

Declared a murderer as his statue was thrown into Bristol harbor in 2020, in the symbolically most potent episode in Britain, Edward Colston (1636–1721) was controversial in death as he was not in life. Television presenters confidently announced as fact that Colston's statue was thrown into the very harbor from which his slaving ships set sail, and that it met a watery grave like the dead and dying slaves thrown from the ships from which he made the bulk of his fortune; but he directly owned no slaving ships, and the bulk of his fortune probably did not derive from the slave trade. In many respects with Colston, we have the problems of addressing many issues for a period in which information is not as full as we would like; not that that prevents commentators.

A child born in Bristol, and fond of the city as a result, Colston left it during the Civil War and was essentially a London merchant. It is unclear how much of his fortune derived from the slave trade, in which he was involved from 1680 to 1692 due to his membership of the Royal African Company of which he was Deputy Governor from 1689 to 1690. Colston was also a partner in a Bristol sugar refinery. In practice, much of his merchant activity was focused on trading with the Mediterranean and Iberia, lucrative trades from which he presumably derived most of his wealth, and Colston was involved with slavery for around one fifth of his long business career. For the last thirty years of his life he was not involved, although, crucially, it is not clear why. It was in that time that he endowed his charities, for education and poor relief, which makes him the greatest philanthropist in Bristol's history.

As far as the general point about memorialization is concerned, it is surely better if matters are handled in a legal and temperate fashion. Feeling strongly about an issue as a justification for mob action could all too readily be used across a society that includes many who feel strongly about other aspects of belief and activity; and then we would be in a very dark place indeed, possibly one of sectarian violence, or of physical attacks on homosexuals or abortion clinics, or a whole range of what is hated by at least someone. I cannot help reflecting on the image of violence in *Sir Thomas More*, a play in the writing of which Shakespeare may have had a role:

And that you sit as kings in your desires,
Authority quite silent by your brawl,
And you in ruff of your opinions clothed;
What had you got? I'll tell you. You had taught
How insolence and strong hand should prevail,
How order should be quelled; and by this pattern
Not one of you should live an aged man,
For other ruffians, as their fancies wrought,
With self same hand, self reasons, and self right,
Would shark on you, and men like ravenous fishes
Would feed on one another.

This passage referred to the ugly May Day 1517 riots in London; riots directed against foreign residents, and the writer vividly refers to refugee foreigners "their babies at their backs."

In 2023, St. Mary Redcliffe Church in Bristol, which in 2020 had removed a stained-glass church window dedicated to Edward Colston, decided, after a competition, to present Jesus in a racially conscious light. One panel depicted Jesus as a black man aboard a British slave ship. Justin Grau, Chancellor of the Diocese of Bristol, in his judgment providing permission for change, stated:

The Church of England and the historical behaviour of this parish church in excusing the life of Colston have a journey of repentance to make. To excuse or ignore the slave trade is a sin. To encourage parishioners to look at a memorial to a slave trader and to be encouraged to "Go thou and do likewise" is not only grotesque but entirely contrary to the Gospel command to love one another.

There is genuine sorrow and contrition, and appropriately so. At the same time, this concern and contrition link to other issues. In part, the discussion of the slave trade and slavery is also an aspect of the longstanding critique of their alleged role in the development of capitalism and Western power.[2] This discussion is an aspect of the pressure for apology that is so prominent in modern public history. Yet this pressure does not only involve Britain, Europe, and the West, as the pressure from China and Korea for an apologetic tone on the part of Japan over its imperialism and military conduct in the early twentieth century indicates. Moreover, in part this demand relates to slavery, notably in the case of the over 100,000 mostly Korean "comfort women"—in practice, enforced and enslaved prostitutes for the Japanese military.

The treatment of the black experience in America, however, is apparently central because of the significance of America in the modern world. Moreover, the black experience is crucial to the discussion of the nature and dynamics of American society. A focus on slavery serves to emphasize separate and distinctive black development, and thus challenges ideas of cultural syncreticism (melding), let alone benign American exceptionalism.[3] The related assertion of African identity by some in the American black rights movement is directly pertinent. William Du Bois (1868–1963), one of the founders, in 1909, of the National Association for the Advancement of Colored People, and, as an historian, the author of *The Suppression of the African Slave Trade* (1896) and the very critical *The World and Africa* (1947), emigrated in 1960 to Ghana where he supported the idea of an *Encyclopaedia Africana*. The assertion of African identity was the case in particular with black separatists, such as Malcolm X, the head of the black Nation of Islam, but also with others who were not separatists.

The winning of independence by black African states proved an inspiration to the Civil Rights movement in Africa, which gathered pace in the wake of Ghana's independence in 1957. Moreover, the American political establishment took note. Having attended the independence celebrations, Vice President Richard Nixon reported to President Dwight Eisenhower:

We cannot talk equality to the peoples of Africa and Asia and practice inequality in the United States. In the national interest, as well as for the moral issues involved, we must support the necessary steps which will assure orderly progress towards the elimination of discrimination in the United States.[4]

Aside from African consciousness, the experience of slavery, and thus the background of the slave trade, served as a reference point in America to those pressing for civil rights. In his "I have a dream" speech, delivered on the steps of the Lincoln Memorial in Washington on August 28, 1963, Martin Luther King Jr. declared that, despite the Emancipation Proclamation of 1863, "One Hundred years later, the life of the Negro is still sadly crippled by the manacles of segregation and the chains of discrimination." The following year, the Civil Rights Act banned employment discrimination on the basis of race, religion, and gender, and decreed an automatic end to the funding of the discriminatory federal program.

The contentious legacy of slavery was accentuated by controversy over civil rights, and, in turn, that provided a perspective for considering slavery and the slave trade. The Civil War (1861–5) proved a key historical issue. In 2000, in response to discussion of the Interior Appropriations Bill, the National Park Service submitted to Congress a report assessing the educational information at Civil War sites and recommending that much be updated, not least to illustrate the "breadth of human experience during the period, and establish the relevance of the war to people today." Representative Jesse Jackson Jr. and other members of Congress had complained that many sites lacked appropriate contextualization and, specifically, that there was often "missing vital information about the role that the institution of slavery played in causing the American Civil War."

The treatment of the Civil War is particularly contentious, with the issue of slavery frequently highlighted in order to criticize anti-bellum (pre-Civil War) Southern culture and to present the South as "unAmerican" or "anti-American." In contrast, Southern apologists emphasize states' rights and play down the role of slavery in Southern separatism in 1861, and misleadingly so.

As throughout the twentieth century, in America, film proved a way to argue the case. A prime instance was presented in 2012 by the much-praised film *Lincoln*, produced by Steven Spielberg. Critically and commercially successful, this film focused on the last stage of the American Civil War in early 1865 and the passage of the Thirteenth Amendment, abolishing slavery. In part, this was an account that was surprising for Americans, as the film showed the political horse-trading by President Lincoln that was necessary for the legislation to pass through the House of Representatives, a theme that, arguably, implicitly criticized President Barack Obama for remaining aloof and not taking part in such horse-trading. The film also focused on the issue of slavery and its consequences as a central topic in American history. This account also provided an implicit means to defend Obama, who, in 2009, became America's first African-American president, and also to present him as the culmination of American history, and thus as a pattern for the future. Obama himself praised the film. Moreover, by depicting Lincoln's opponents critically, a harsh light was cast, by implication, on Obama's opponents. That Lincoln was a Republican, as are Obama's opponents, served to indicate that that party had changed its role, and could make it appear unworthy of its heritage.

Thus, controlling and defining the American past, and notably with regard to slavery, becomes an aspect of current politics in America. There were apologies for American history, as by the Senate in 2005 for neglecting to pass legislation in the late nineteenth and early twentieth centuries to make lynching a federal crime. However, charges of exploitation and of historic wrongs explaining present circumstances are contested, especially from the "white South." The latter, indeed, has its own aggressive and self-righteous sense of historical grievance. In 1990, Virginia elected America's first black governor, Douglas Wilder, a grandson of slaves. At the same time, there was considerable resistance in parts of the South—for example, Mississippi—to abandon the Confederate flag, and other symbols of difference and defiance. To critics, these symbols contributed to the intimidation of black people.

In 1998, David Beasley lost his post as Governor of South Carolina, a state where symbolism played a continuing role, for supporting the removal of the flag from the statehouse.

A victim of revolution, George III's statue has long gone from New York. Now that of his most bitter critic, Thomas Jefferson, is to follow, being no longer acceptable in the City Hall of a beacon that for long has represented the United States. Not only the writer of the Declaration of Independence, and a major figure in Virginia's war, Jefferson was also a two-term President. His public service included being Ambassador to France.

The history wars in America are vigorous, and in some respects longer-lasting than those in Britain, in which monarchy in the absence, since the mid-eighteenth century, of division over dynasty, provides a basic harmony that is absent from a country still digesting in its public history a more recent civil war as well as confronting every four years the issues bound up in the election of a head of state. The latter make the reputation of past presidents particularly contentious, and Jefferson attracts ire from those seeking to direct attention against the national origin-account.

Slavery has been used to weaponize not so much the conflict between Left and Right, although that plays a role, as that within the Left. It is akin therefore to that in Britain. Pushing the Left leftwards offers possibilities for power, or at least influence, to politicians and commentators who wish to direct the Democrat Party/Labour, and thereby to gain control of the state. The last appears even more attractive as government powers have expanded in the age of Covid.

That may appear a fanciful reading of the departure of statues, but the latter is but the most visual exemplar of a more wide-ranging process, from school curricula to street names, designed to banish an established sense of identity. We are in part seeing a process comparable to the "Cultural Revolutions" attempted from China and Cambodia to campuses, from the 1960s on. In the case of America, history is understandably to the fore, because America, unlike all other major states, has an historic constitution, unlike those dating from the twentieth century or (as with Britain leaving the EU) more recently. That makes intentions and contexts of the Founding Fathers particularly contentious today. The assault was focused by the 1619 Project, launched in the *New York Times* in 2019, which "aims to reframe the country's history by placing the consequences of slavery and the contributions of Black Americans at the very center of the United States' national narrative."

The contentious accuracy of some of the claims for example the unfounded assertion the colonists sought independence to protect slavery, are beside the point, as the intention is to assert a new "truth" that condemns modern America by reference to this legacy. This condemnation rests in part on an assault on the alternative narrative linked to the American Revolution, as well as a downplaying of later white people such as Lincoln associated with the struggle for Civil Rights.

Controversies are exacerbated when other issues, such as gender, are involved. Thus, the charge that Thomas Jefferson had had an affair with an African-American servant, Sally Heming, led to contention in the 1990s and 2000s. The charge was seen by some as an assault on the integrity of the Founding Fathers. In April 2014, the issue of slavery was added to defiance of the Bureau of Land Management in America's West, when Cliven Bundy, a prominent opponent of the bureau, allegedly claimed that African-Americans had been well off as slaves.[5]

There is also the issue of how visitors to some former plantations worked by slaves were (and still are) presented a misleading account. A British journalist of Indian descent observed of the American South in 2014: "Even more startling than the casual omission of the whipping, executions, shackling, rape, mutilation, branding, burning and imprisonment that went on was the nostalgic aggrandizement of the Deep South's history."[6] At the same time, the slave experience has been recently highlighted at prominent sites, including Monticello and Montpellier, the homes of Presidents Jefferson and Madison respectively.

In America, as in Brazil, the poor remain, to this day, disproportionately black, and black people are disproportionately poor. The percentage of black people below the poverty line in America was 35.7 in 1983, 22.5 in 2000, and 24.7 in 2004, compared to percentages for non-Hispanic white people of 8–9. At the national level, in America in the 2000s, black mothers were twice as likely as white counterparts to give birth to a low-weight baby, and their children were twice as likely to die before their first birthday. Furthermore, black people were disproportionately numerous among the imprisoned. In Brazil, slave-like conditions for Native Americans, in which debt bondage is linked to the seizure of identity papers, and to cruel treatment by armed guards, are seen today—for example, in the pig-iron industry round Maraba.

A very different and longstanding legacy was that of the use of opposition to the slave trade and slavery as a call to encourage action in other respects, a process that continues to this day. For example, writing an opinion piece in the *Times* on March 8, 2014, Mariella Frostrup acclaimed British legislation establishing gender equality as a criterion for development aid, the first state to make that criterion a legal condition:

> The impact of putting gender equality at the heart of development aid has a similar potential to influence global policy as William Wilberforce's Slave Trade Act of 1807—a historic piece of legislation that set a precedent for other nations to follow.[7]

As in America, with the Civil War and civil rights, slavery and the slave trade were, and are, intertwined as issues with other aspects of public politics and history.

In the case of states where slavery was common, there is frequently an attempt to argue that pressure from below—in other words, from the slaves themselves—was important in bringing about Abolition. This is a particularly significant argument for colonies that have gained independence. In Brazil, there is an emphasis on pressure from below as well as from civil society pressure. This argument is different in emphasis from that which focuses on the weakening of slavery due to economic pressures—notably, the end in 1850 due to British pressure of the slave trade to Brazil which drove up the price of slaves by preventing fresh flows of captives, such that slavery ceased to be considered fundamental for the future of Brazil. Deep-seated regional and urban–rural rivalries were exacerbated by this development. Thus, far from there being solely one factor, a number, including moral pressure, were involved as the forces favoring Abolition greatly outstripped those favoring the survival of slavery.

The situation among former slave-trading states varied and varies. Alongside countries where the legacy is prominent and contentious, it is important to note others. The slave trade and slavery were, and are, for example, a less prominent issue in the Netherlands and, even more, Denmark than in Britain. Moreover, alongside apologies and pressure for remorse and compensation on the part of the countries from which slave-traders came, it is important to give due weight to African agency in the trade. This agency was longstanding, with brokers and traders crucial not only as intermediaries but also in shaping the trade.[8] The descendants of those who provided and sold slaves to Western traders have not felt called upon to assess their role. This role does not make the responsibility of these traders less significant, but it is pertinent. A stress on the active role of Africans emerges more clearly as the West's relative military and political weakness in Africa prior to the late nineteenth century is considered.

This is one of the themes of my book, one that arises from my work on military history, and that helps to explain why that is emphasized in the book. Attention should be directed, as in this book, to a subject that is obscure due to the nature of the sources—namely, conflict between Africans. This warfare resulted in numerous captives, whether or not they were slaughtered or kept as slaves. If the latter, some would be retained for slavery by their captors, a subject that receives insufficient attention, and some would be sold for distant slavery, again both in Africa and further afield. The latter entailed both the Western and the Arab slave trades. The Arab role went on after the Westerners had ceased their part. This role can be amplified by considering the continuing ethnic tensions that underline the very harsh treatment of black people in

Darfur in West Sudan in the 2000s, 2010s and 2020s, and in southern Sudan before it gained independence from Sudan in 2011.

This more widespread, even apparently general, aspect of the slave trade and slavery invites the reflection that Atlantic slavery was exceptional not because of the slave trade, but because this slavery led to the racism that is of lasting importance in defining white–black relations. A distinctive presentation of blackness can be seen from Classical Rome.[9] Nevertheless, Atlantic slavery has been much more important for recent attitudes.

The Slave Trade Today

The recent and current large-scale movement of Africans to Europe, particularly Spain and Italy, offers an instructive comparison and contrast. On the Atlantic route to Spain, most came from West Africa—for example, from Mali and Burkina Faso—and seek work in Europe. A key means of access is from Mauritania, Gambia, and, mostly, Senegal, by boat to the Canary Islands, a Spanish possession where they seek asylum. There is a similar movement from North Africa to Italy, notably the islands to the south of Sicily. The chaos in Libya after the overthrow of the Gaddafi government in 2011 meant that Italian attempts to persuade Libya to deter emigration were no longer effective: until 2011 the immigrants had been blocked at sea and many were returned to Libya. Crossing in open boats, the Africans are exposed to the sun and generally short of water. The boats are very crowded and, aside from the danger of shipwreck, a certain number die in the crossing.

The relationship between this migration and the slave trade is at best indirect, as the would-be migrants are not slaves, while the entrepreneurs who organized the traffic were mostly African. Looked at differently, however, if the emphasis is on economics, the contemporary example underlines the extent to which labor flows are frequently harsh. It also indicates that the combined pressure of large-scale demographic expansion and serious economic difficulties ensures that labor is now exported from Africa. Slavery is not part of the process, although once arrived in Europe or North America, immigrants, notably if illegal, are frequently seriously exploited. Indeed, as a result, in one light, this is, in fact, slavery without the "protection cost" of having to provide for and control the slaves.

A pertinent comparison that helps explain why Africans flee Africa today may be not to the classical Atlantic slave trade that ended in the nineteenth century, but rather to the impact on African economies in the nineteenth century of the arrival of Western goods, particularly textiles and metals. These goods had already been attractive in the eighteenth century, but, from the coming of peace in Europe in 1815 and even more from the mid-nineteenth century, became available in larger and cheaper quantities thanks to the economies of scale brought by industrial development in Europe and the use of steamships. Protectionism in European markets encouraged exports to Africa. These goods hit African industries hard. The results included the growth in Africa of primary production focused on the Western market—for example, of palm oil—and a rise in the slave trade within Africa, for example to support the major textile industry in Kano. This rise was exploited for the last stages of the Atlantic slave trade as well as for the Arab slave trade.

The sight of Africans and others—for example, Sri Lankans—in open boats being intercepted by the Spanish, Italian, and Australian navies in the 2000s and 2010s in an attempt to keep them from the European and Australian economies invites attention to the varied relationships, across time, between globalization, movement for work, and political flight, with a coda about the different role of naval power today compared to that of the nineteenth-century British navy. The current situation also underlines the continuing complexity of the relationship between Africa and the forces in the world economy.

The same point is pertinent for labor flows elsewhere and not involving Africa—for example, from Latin America to America—although, as from Africa, many of those who travel are refugees as well as work-seekers. Some of those who travel north are seriously exploited, both as they travel and once they have arrived. The exploiters include both those who transport the would-be migrants and those for whom they subsequently work. The same is true of the movement from the Balkans, and elsewhere, into the European Union. People from outside the latter are frequently trafficked in, becoming illegal immigrants. Many are then used for work in which they are treated harshly and paid very little, if anything.[10]

Prostitution is a major element of this trade, and, although some women travel voluntarily, others are frequently brutalized as part of the trafficking. This is a reminder of the violence, degradation, and misery that is part of the slave trade. Moreover, there is a specific sexual dimension, with modern sex slavery being a continuation of the sexual victimization of enslaved people in previous ages. A related form of exploitation is provided by the extent to which those vulnerable to slavery disproportionately included and include orphaned children.[11] Technology enters into the issue with the spread from the 2010s of "virtual trafficking," where predators exploit children in video-chat rooms.

Aside from international trafficking, there is also trafficking within countries. Thus, in Britain, as elsewhere, the homeless are trafficked for work on exploitative terms, "rip-off work," in what is seen as new slavery. The homeless come forward to seek work, but they are definitely exploited.[12]

The exploitation seen in Europe can also be found elsewhere. Thus, North Koreans who have fled into China are frequently trafficked: women are used as prostitutes and shackled when not working so that they cannot escape. States that have very different political systems also see trafficking. In 2014, there were revelations about the use of slave labor to produce prawns in Thailand, a democracy, albeit one that was overthrown by a military coup that year. In 2014, the US State Department relegated Thailand to the lowest rank in its Trafficking in Persons report, which meant that it was ranked alongside North Korea, Iran, and Saudi Arabia in the way it protected workers from abuse.[13]

Slavery today[14] is a brutal part of an often wider range of harsh labor conditions. For the Gulf, the key source of labor is no longer slaves from East Africa, but nominally free workers from South Asia, who are subject to harsh controls. For example, several hundred thousand Nepalese construction workers, brought into Qatar in the 2010s to build its facilities and infrastructure for the 2022 World Cup, experienced not only 12-hour shifts, sweltering heat, and often hazardous environments, resulting in high death rates, but also labor controls in the shape of the confiscation of the passports of migrant workers, a frequent situation. Pay is often withheld for months. The condition of workers in the Gulf does not amount to the slavery that can be seen elsewhere in the world. However, it represents a form of indentured servitude that scarcely matches free labor conditions. Similar points can be made about the women brought into the Gulf to act as servants. A particularly important aspect is the use of women as maids. Some have been mistreated. Qatar's 1.39 million migrant workers formed 94 percent of the workforce in 2013. The sponsorship system, *kafala*, gives employers great control. Exit permits have to be approved by employers, the excuse being that, otherwise, employers would be liable for debts and criminal activity. Moreover, the ultimate sanction for breach of contract is criminal detention. According to a 2014 report by N.A. Piper, a law firm commissioned by Qatar's government, this detention "lends weight to the allegation that the *kafala* system constitutes forced labour."[15]

A very different form of labor that may in the future be discussed in terms of slavery relates to the enhancement of robots and the addition of human characteristics. Herbert Simon, a pioneer in artificial intelligence (AI), claimed in 1964 that machines would be capable within 20 years of doing any work a human could do. These high hopes proved mistaken, as often happens with technological futurology, but the work that could be done by machines greatly expanded, while in 2023 there was a strong revival in the discussion of AI. Claims that machine-work might lead to a new form of slavery, however, were misplaced, as machines are not human. Instead, the comparison is more clearly with the use of animals to provide labor.

The extent to which the situation might change if cloning extended to humans invites speculation. A sense that modernity brought unwelcome power was captured in the arts, notably, the modernization of the horror genre with a concern about zombies, the living dead, as in George Romero's films—for example, *Land of the Dead* (2005). Moreover, the idea of clones raised underground to provide perfect body parts for their donors was the theme of the film *The Island* (2005).

Such stories captured the idea that a new technology would enable humans to develop a brutalized subhuman world, and also the anxiety that such a world might be the fate of the species itself. In short, a new, more insistent slavery was outlined in the world of fiction. Linked to this were concerns about the nature of the society that might be responsible for such an outcome, concerns that refracted anxieties about changes and possible future developments in the world.

Reparations

Reparations became a more significant issue in the 2020s, but it is also unclear who reparations should be paid to, in what forms, and whether even the acceptance of money would be a tacit form of endorsement.[16] In practice, the emphasis is on the group, not the individual, and the relevant identity politics focuses on an idea of hereditary hardship. In March 2014, Caricom, the fifteen-strong Caribbean Community of states, demanded reparations from European slaving nations for what it presented as the enduring legacy of the slave trade, including "psychological trauma." Sir Hilary Beckles, a historian from Barbados who chaired the Caricom Reparations Commission, argued that year:

This is about the persistent harm and suffering experienced today by the descendants of slavery and genocide that is the primary cause of development failure in the Caribbean … The African descended population in the Caribbean has the highest incidence in the world of the chronic diseases hypertension and type 2 diabetes, a direct result of the diet, physical and emotional brutality and overall stress associated with slavery, genocide and apartheid … The British in particular left the black and indigenous communities in a general state of illiteracy and 70 per cent of blacks in British colonies were functionally illiterate in the 1960s when nation states began to appear.

This running together of slavery and imperial rule confused differing phenomena. Both Singapore and South Korea (a Japanese colony) indicate that the latter did not necessarily lead to long-term poverty. However, imperial rule as an issue does have the merit of directing attention to "public slavery," the facet that tends to attract very little scrutiny today. The focus instead is on "private slavery," in the sense of ownership by individuals or corporations, and as part of a critique of capitalism. However, as a global account of slavery this is inadequate. Being a slave of the state is a characteristic that bridges from Antiquity to the present, and, again, that raises the problematic nature of compensation.

So also with the concept of hereditary guilt. The most prominent instance of this pernicious idea was the anti-Semitic practice of blaming the Jews, all Jews, and forever, for the Roman execution of a Jew. This practice has long been castigated, which makes it ironic to see hereditary guilt so much to the fore in modern identity politics. These politics are about competitive anger and "slavery" provides the entry point, narrative and anger for those of African descent. However, the cruelty and misery of conflict in some parts of Africa over the last two decades, notably Sudan, South Sudan, the Central African Republic, Congo, Nigeria, and Mali, indicate that Western pressure and control were scarcely necessary features. When warfare in Africa, past or present, is discussed, it is often attributed to the pressures and opportunities of European slave trading (the Arabs, slavers on a massive scale, tend to get a free pass, but this warfare was not necessarily dependent on these opportunities. For example, the scale and bitterness of the *Mfecane* wars in southern Africa, caused by the rise of the Zulu empire in the early nineteenth century and by competition for resources, cannot be readily linked to the slave trade, nor indeed attributed to exogenous pressure in the shape of European actions. As with the case of the provision of weapons in modern conflicts, such pressures did play a role, but they were instrumental rather than causative.

Critical views of the slave trade and slavery are not lessened by the large degree of African, Arab, and other agency, but it certainly complicates any blame game, and, indeed, should, rather, direct attention to slavery in the world today, whether public (North Korea), ethnic (Indian caste system), gender (mistreatment of women in many countries), or economic. Estimates of the number of slaves in the world today vary, but 20 million is an oft-cited figure. Pursuing blame and reparations for the situation 200 years ago or longer to a degree is a distraction from this issue, and one that invites consideration about motives and methods.

For America, the argument that Civil Rights are incomplete without reparations ignores a host of factors, not least the extent of white poverty and the practical problems, institutional and legal, of formulating and implementing any such policy.

The definition of slavery in terms of partisan purposes is a key element. The emphasis on race and capitalism is part of the major success by the Left in recent years in advancing its intellectual agenda. The increased willingness of the young to define themselves as leftwing is part of the equation, with a key element its ethnic, gender and sexual orientation politicizing. With the Left unable to use the Holocaust, a key signifier in historical consciousness, they have focused on the slave trade as an alternative and an equivalent, the latter bizarre as the purpose was not to kill slaves.

Conclusion

Slavery and the slave trade were, and are, vile. It is good that they no longer exist in the form and at the scale discussed in this book. They are vile, but not unique as forms of labor control nor as reflections of ethnic and other prejudice.

Similarly, the vileness of the Atlantic slave trade is not lessened by the large degree of African agency nor by the equivalent Arab slave trade. Yet, these important elements serve as a reminder of the need for caution before moving from analysis to judgment.

Slavery and the slave trade persist because of the points brought up in the text about its historical place, in terms of economic, social, and political factors, as well as racism and the related detachment from the "other." On some level, such harshness and prejudice suggest that part of us has not changed, may perhaps never change, and needs guarding against. The human damage, if not wreckage, that arises from harshness and prejudice is all too apparent for the slave trade of the past, but is also the case of the present.

The curse of the past is not what happened but rather the inability to build on the past when looking at the present and to the future. The latter, however, should not be confronted in terms rigidly dictated by a sense of historic grievance, a sense that can lead to the denial of choice and the related neglect of opportunities. The use of a public account of past suffering as a justification for particular political and cultural strategies in the present is unhelpful psychologically as well as divisive politically, and frequently misleading.

In Shakespeare's play *Romeo and Juliet* (*c.*1595), a distraught Romeo tells Friar Lawrence, who is trying to offer reasonable counsel: "Thou canst not speak of that thou dost not feel." There certainly are too many today who suffer the pain and humiliation of slavery. Estimates vary, but the figures usually range from 20 to 30 million. Looking to the future involves considering their plight as well as recalling those enslaved in the past.

Notes

1 On which see J. Black, *Imperial Legacies: The British Empire Around the World* (New York, 2019).
2 E. Williams, *Capitalism and Slavery* (Chapel Hill, NC, 1944); C. Hall et al., *Legacies of British Slave-ownership: Colonial Slavery and the Formation of Victorian Britain* (Cambridge, 2016); K. Andrews, *The New Age of Empire: How Racism and Colonialism Still Rule the World* (London, 2021).
3 S.W. Mintz and R. Price, *The Birth of African American Culture: An Anthropological Perspective* (Boston, MA, 1992).
4 Quoted in C. Fraser, "Reframing Freedom and Citizenship in the Black Atlantic: MLK Jr., Ghana's Independence, and the Shifting Terrain of History in the Atlantic World," in *Africa, Empire and Globalization*, eds. T. Falola and E. Brownell (Durham, NC, 2011), 512.

5 *The Times*, April 26, 2014.
6 S. Sanghere, "Just Don't Mention the Slaves: My Surreal History in America's Deep South," *The Times*, March 4, 2014, 6.
7 M. Frostrup, "Britain Shows the World the Way—Again," *The Times*, March 8, 2014, 26.
8 T. Green, ed., *Brokers of Change: Atlantic Commerce and Cultures in Pre-Colonial Western Africa* (Oxford, 2012).
9 F.M. Snowden, *Before Color Prejudice: An Ancient View of Blacks* (Cambridge, MA, 1983).
10 J. Salt and J. Stein, "Migration as a Business: The Case of Trafficking," *International Migration*, 35 (1997), 467–91.
11 G. Campbell, S. Miers, and J.C. Miller, eds., *Children in Slavery through the Ages* (Athens, OH, 2009).
12 *File on Four*, BBC Radio 4, May 13, 2014.
13 *The Guardian*, June 25, 2014, p. 2.
14 K. Bales, *Disposable People: New Slavery in the Global Economy* (Berkeley, CA, 1999).
15 *Financial Times*, May 15, 2014, 3.
16 As argued in J. Sweet, "Slave Trading as a Corporate Criminal Conspiracy, from the Calabar Massacre to BLM, 1767–2022," *American Historical Review*, 128 (2023), 29.

Selected Further Reading

The emphasis is on recent works, as earlier works can be followed up through their footnotes and bibliographies. In line with the readership, the emphasis here is on English-language works. A searchable online version of an excellent bibliography prepared by Joseph Miller can be found at www2.vcdh. virginia.edu/bib/. Another key database is *Voyages: The Trans-Atlantic Slave Trade Database* (www.slavevoyages.org).

Alencastro, L.F. de., *The Trade in the Living: The Formation of Brazil in the South Atlantic, Sixteenth to Seventeenth Centuries* (2019).
Bales, K., *Understanding Global Slavery* (2005).
Baucom, I., *Specters of the Atlantic: Finance Capital, Slavery, and the Philosophy of History* (2005).
Brown, C.L., *Moral Capital: Foundations of British Abolitionism* (2006).
Byrd, A.X., *Captives and Voyagers: Black Migrants across the Eighteenth-Century British Atlantic World* (2008).
Candido, M.P., *An African Slaving Port and the Atlantic World: Benguela and Its Hinterland* (2013).
Chachem, M.W., *The Old Regime and the Haitian Revolution* (2012).
Christopher, E., *Slave Ship Sailors and their Captive Cargoes* (2006).
Curtin, P., *The Slave Trade: A Census* (1969).
Davis, D.B., *Inhuman Bondage: The Rise and Fall of Slavery in the New World* (2006).
Davis, R.C., *Christian Slaves, Muslim Masters: White Slavery in the Mediterranean, the Barbary Coast, and Italy, 1500–1800* (2004).
Diptee, A., *From Africa to Jamaica: The Making of an Atlantic Slave Society, 1775–1807* (2010).
Drescher, S., *Abolition. A History of Slavery and Antislavery* (2009).
Eltis, D., ed., *Coerced and Free Migration: Global Perspectives* (2002).
Eltis, D. and Engerman, S.L., eds., *The Cambridge World History of Slavery* (2011).
Eltis, D. and Richardson, D., *Extending the Frontiers: Essays on the New Transatlantic Slave Trade Database* (2008).
Eltis, D. and Richardson, D., *Atlas of the Transatlantic Slave Trade* (2010).
Engerman, S.L., *Slavery, Emancipation, and Freedom: Comparative Perspectives* (2007).

Ferreira, R., *Cross-Cultural Exchange in the Atlantic World: Angola and Brazil during the Era of the Slave Trade* (2012).

Gomez, M.A., *Reversing Sail: A History of the African Diaspora* (2nd ed., 2019).

Green, T., *The Rise of the Trans-Atlantic Slave Trade in Western Africa, 1300–1589* (2012).

Hamilton, K.A. and Salmon, P., eds., *Slavery, Diplomacy and Empire: Britain and the Suppression of the Slave Trade, 1807–1975* (2009).

Harris, R., *The Diligent: A Voyage through the Worlds of the Slave Trade* (2003).

Hawthorne, W., *From Africa to Brazil: Culture, Identity, and an Atlantic Slave Trade, 1600–1830* (2010).

Hochschild, A., *Bury the Chains: Prophets and Rebels in the Fight to Free an Empire's Slaves* (2005).

Huzzey, R., *Freedom Burning: Anti-Slavery and Empire in Victorian Britain* (2012).

Inikori, J.E. and Engerman, S.L., eds, *The Atlantic Slave Trade: Effects on Economies, Societies and Peoples in Africa, the Americas, and Europe* (1992).

Klein, H.S., *The Atlantic Slave Trade* (2nd ed., 2010).

Law, R. and Lovejoy, P.E., *The Biography of Mahommah Gardo Baquaqua: His Passage from Slavery to Freedom in Africa and America* (2002).

Lovejoy, H., ed., *Liberated Africans and the Abolition of the Slave Trade, 1807–1896* (2020).

Lovejoy, H., ed., *Regenerated Identities: Documenting African Lives* (2022).

Manning, P., *Slavery and African Life: Occidental, Oriental, and African Slave Trades* (1990).

Menard, R.R., *Sweet Negotiations: Sugar, Slavery, and Plantation Agriculture in Early Barbados* (2006).

Nellis, E., *Shaping the New World: African Slavery in the Americas, 1500–1888* (2013).

Nwokeji, G.U., *The Slave Trade and Culture in the Bight of Biafra: An African Society in the Atlantic World* (2010).

Paquette, R.L. and Smith, M.W., eds., *The Oxford Handbook of Slavery in the Americas* (2010).

Rediker, M., *The Slave Ship: A Human History* (2007).

Rushforth, B., *Bonds of Alliance: Indigenous and Atlantic Slaveries in New France* (2012).

St Clair, W., *The Door of No Return: The History of Cape Coast Castle and the Trans-atlantic Slave Trade* (2006).

Schama, S., *Rough Crossings: Britain, the Slaves and the American Revolution* (2006).

Schumway, R., *The Fante and the Transatlantic Slave Trade* (2011).

Sparks, R.J., *The Two Princes of Calabar: An Eighteenth-Century Atlantic Odyssey* (2004).

Spear, J.M., *Race, Sex, and Social Order in Early New Orleans* (2009).

Thornton, J., *Africa and Africans in the Making of the Atlantic World, 1400–1800* (1998).

Tomich, D. and Lovejoy, P.E., eds, *The Atlantic and Africa: The Second Slavery and Beyond* (2021).

Walvin, J., *A Short History of Slavery* (2007).

Walvin, J., *Crossings: Africa, the Americas and the Atlantic Slave Trade* (2013).

Zacek, N.A., *Settler Society in the English Leeward Islands, 1670–1776* (2010).

Index

Pages followed by "n" refer to notes.

Abd al-Kadir 66
Abolition Act 1807 131, 162
Abolitionism 119; Anglo-American 104;
    in Brazil 139; in Britain 120–5,
    129–30, 133, 146–7; in France 126,
    150; and rhetoric of liberty 7–8; and
    serfdom 3; and Sierra Leone 168;
    and strategies of betterment 142; in
    United States 137; and *Zong*
    massacre 75
abstention movement 124
Accadia 80
Accra 30, 36, 54, 144
Africa: and Atlantic slave trade 97–101,
    121; British exports to 92–3; and end
    of slave trade 143–5, 165–9;
    exploration of interior 39–40,
    99–100; history of slavery in 13–4,
    65; labor shortages in 34; large-scale
    migration from 182–3; resistance to
    slavery in 101–4; social impact of
    slave trade 64–5, 98, 109; warfare in
    62, 65–6, 97–8, 181, 186; Western
    expansion into 18–9; Western
    response to 28; *see also* East Africa;
    sub-Saharan Africa; West Africa
African agency 55, 60, 70, 98, 181,
    186–7
African economies 64, 101, 182
African elites 54–5, 100
African identity 177
African labor, value of 60–1
African slave trades 2; Bristol and 11;
    and internal conflict 35–6; Portugal
    and 24; range of 13–6

African slaves 41; demand for 29–30; in
    Islamic world 15; in Latin America
    26–9; in North America 57, 105;
    society of 33; supply of 30–7;
    trauma and pain of 66–9
African states, black 177
African Trade Act 1750 96
Age of Discovery 21
Age of Revolution 21
agriculture: African 61; large-scale
    systems 9; plantation 15, 26, 29, 44–5
AI (artificial intelligence) 184
Albinus, Bernardo 108
Alcazarquivir 14, 37
Aleuts 29
Alexander the Great 10
Alexander VI, Pope 36
Alfonso III of Aragon 18
Algiers 14, 27, 140
American Civil War 128, 136, 151, 163;
    historical issue of 177–8
American Colonization Society 168
American Revolution 7, 86, 102,
    111–12, 131; historical narrative of
    179; impact on Britain 122–3, 129
American South 3, 149, 156; attempts
    to extend slavery 136–8; and Britain
    149; cotton in 87; historical memory
    of 178, 180; secession of 128, 151–2;
    segregation in 164; *see also*
    Confederate States of America
Americas: end of slavery in 17; force
    against native peoples of 2, 24–5;
    need for labor in 15–6, 23, 25–7; *see
    also* Brazil; North America; Latin

America; Spanish America; United States of America
Amsterdam 80
Anglo-Dutch Wars 54
Angola: and Atlantic slave trade 30, 47, 87, 89, 135; disease in 35; plantation economy in 165; Portuguese wars in 61–2
Annobon 87
Antigua 46, 49, 91, 102
antisemitism xi
Anti-Slavery Commissioner x
apologies for slavery 13, 174, 176, 178, 181
apprenticeship 5, 162, 168
Arab slavers 64, 100, 167, 181, 183, 186–7
Arabia, slaves in contemporary 166
Aragon 18
Argentina 90, 138
Aristotle 9, 28
Arnold, Nicholas 37
art, African slaves in 32
Asante 62–3, 97, 144
asientos 27, 79–80, 89
Assinie 49, 82
Association for Promoting the Discovery of the Interior Parts of Africa 99–100
Atlantic economy 33, 38, 45, 64–5
Atlantic slave trade 1–2, 14, 187; and African slavery 60–3; after British abolition 134–5; Brazil and 27, 32–3, 48, 54, 87; and commercial economy 8, 95; conflicts over 84–6; end of 119, 145–6, 165; height of 76, 79; international competition in 55–6; and labor flows 16, 41; origin of 23; and plantation crops 44–5; prior to 1600 28; and racism 182; specialization of 89; and Western development 21
Atlantic world viii
Augustine of Hippo, Saint 10
Auschwitz 171
Australia 4
autonomy 70, 103
Axim 30, 36, 80
Azores 32
Aztecs 24

Baghdad 15
Bahamas 50, 107, 112
Bahia 89, 146
Balkans 12, 15, 183
Bambara 106
Barbados: African slavery in 49–51; after slavery 106; and Carolina 58–9; indentured labor in 4; in literature 109; and Panama Canal 163; slave protests in 103; slave rebellions in 69, 146; SPG in 108; sugar in 46
Barbados Slave Act 51–2, 58
Barbary states 140
Barbuda 70
Barfleur 56
Barnstaple 92
Barrère, Pierre 108
Beasley, David 179
Beckles, Hilary 185
Belize 84
Benezet, Anthony 104
Benguela 47
Benin Coast 53, 82
Berbice 145
Bermuda 12, 49, 91, 106–7
Bible 40
Biddle, James 143
Bideford 92
Bight of Benin 54, 135
Bight of Biafra 54, 60, 106, 135, 141
Biohó, Benkos 21
Bissagos Islands 21, 35
Black America 69–71, 177
Black churches 71
Black Death 23
black people 71, 104; after American Civil War 154; and American Revolution 112; in modern United States 177–80; racist beliefs about 107–11, 126; in Spanish America 30; in Sudan 181–2; and tropical diseases 57; in West Indies 50, 82; *see also* free blacks
Black Sea 10, 18, 23–4, 47
black soldiers: in American Civil War 112, 154, 163; in American Revolution 111; in Caribbean 128; in French Guiana 90; in Seminole Wars 146; in Spanish Florida 102; in War of 1812 133

Blackness 40–1, 110, 182
Blair, Tony 174
Blumenbach, Johann Friedrich 110–11
bodies of slaves 51
Bogle, Paul 165
Boko Haram 169
Bolivia 26, 29
Bombay 134
Bonny 119, 135
Bordeaux 44, 56, 83, 94
Borno state 169
Bornu 14
Botswana 145
Brandenburg-Prussia 55, 80
branding 150, 180
Brazil ix, 1, 15; and Angola 88–9, 133; and Atlantic slave trade 27–33, 47–8, 58, 88; black and mulatto soldiers in 90; cacao in 46; contemporary condition of black and Native peoples 180; end of slavery in 151, 155–7, 165, 181; ending slave trade 133, 138–9, 141, 143; internal slave trade 136; north-east 32–3, 47, 87, 89, 94; northern 38; Portuguese conquest of 25; slave economy of 29–30, 32–3, 47–8, 134–5; slave life in 106, 147; slave rebellions in 26, 33, 129, 146; south of 87–8
Breckinridge, John 152
Brissot, Jacques-Pierre 126
Bristol 11, 44, 90–2, 96; legacy of slavery in 175–6
Britain 1; abolition of slave trade 120–6, 129–33; deporting convicts from 4; expansion in Africa 61–2; and Haitian revolution 128; industrialization of 113; naval action against slave trade 138–43, 167, 181; and Portugal 48; public discourse on slavery 3, 7; and the slave trade 79–80, 86, 90–7; *see also* England; Scotland
British Atlantic 48, 92, 107, 113
British commercial system 88, 94, 96
British East India Company 84, 92–3
British Empire: after slavery 162, 167; deaths of slaves in 68; end of slavery in 147–9; responsibility for slavery 175, 185
British Guiana *see* Guyana

British Royal African Colonial Corps 144
British West Indies 49, 91; after end of slave trade 145–6; after slavery 163–5; and North American colonies 58, 86, 94; slave life in 106; slave risings in 102; sugar from 87, 96, 129
brotherhoods 71, 110–11
Browne, William 100
buccaneering 50
Buchanan, James 152
bullion 29, 88
Bundy, Cliven 180
Burgess, Thomas 123, 130
Bush, George W. 79
Bush Negroes 103
Bute, Earl of 86
Butri 80
Buxton, Thomas 142

Cabinda 135
cacao 45–6, 51, 83, 89–90
Caesar, Julius 9
caffeine drinks 44–5
California 151
Cameroon 169
Canary Islands 23–4, 26, 30, 32, 182
Candide (Voltaire) 76, 78, 104
cannibalism 66
canoe construction 70
Cantino map 39
Cão, Diogo 36
Cape Coast Castle 54, 63, 67, 144
Cape Colony 80, 99–100, 126
Cape of Good Hope 22, 36
Cape Squadron 167
Cape Town 63, 80, 126, 141
Cape Verde Islands 32
capitalism 3, 83, 176, 185, 187
Capuchin Missions 107
Caracas Company 83, 103
Caribs 49
Caribbean *see* West Indies
Caricom viii, 185
Carolina: in American War of Independence 86; establishment of 58–9; *see also* North Carolina; South Carolina
Carolina Slave Act 68
carvel building 39
cash crops 29, 162

cassava 60
caste systems 26, 186
Castile 18–9, 26
castration 68–9
Catholic Church 71, 101
Catholic emancipation 131
cattle herding 70
Cayenne 38, 48, 90, 132
Cayman Islands 50
Central African Republic 66, 186
Ceuta 18
Chad 169
Charles I of England 53
Charles II of England 53
Charles V, Emperor 38
Charles X of France 150
Charleston 94
Chesapeake states 57–8, 91, 105–6, 136
Chevalier de Jaucourt 113
children: biracial 59; illegitimate 106;
    slave 112; trafficking of 183
Chile 25, 90, 138
China: slavery in 4; tea trade with 22, 88
chocolate 44–5, 82
Christian slaves 13–4, 59, 78, 140,
    147, 169
Christianity: and anti-slavery 113, 123,
    130, 149; and French Revolution
    119–20; and Portuguese empire 36;
    and slavery 11–3, 69, 79, 107–8, 126;
    and tropical lands 28
Christiansborg 63, 81
Church Missionary Society 168
Church of England 107, 123, 174, 176
civil rights 150, 177, 181
Civil Rights Act 1866 154
Civil Rights Act 1964 177
civilizing mission 169
Clapham Sect 133
Clarkson, Thomas 104, 123
cloves 166
Cocofio 103
Code Noir 52, 60
Codrington plantation 108
coercion 2
coffee 22, 44, 78; in Brazil 136, 139; and
    demand for slavery 134; European
    consumption of 82–4; French
    sources of 82
Colleton, John 58
Colman, George the Younger 109

Colomb, Philip 167
Colombia 21, 25, 85
colonialism, and end of slavery 166
colonization 12, 24, 27, 29, 44
Colston, Edward 175–6
Columbus, Christopher 21, 23, 39
commercial economy 8, 92, 166
Commercial Revolution 80
Committee for the Relief of the Black
    Poor 168
commodification 17, 50, 68
Company of Merchants Trading to
    Africa 96, 144
Company of Royal Adventurers
    Trading into Africa 53, 56; *see also*
    Royal African Company
compass, magnetic 39
concentration camps 171
concubinage 3
Confederate States of America 151,
    153–4, 163; flag of 178–9
Congo 66, 97, 107, 186
Congo (steamship) 144
Congo River 135, 144
Congress of Berlin 169
Congress of Vienna 133, 145
Constantinople 5, 23, 59
consumerism, popular 70
conucos 70
convicts: enslavement of 10, 78;
    transporting 4, 41
Coriso Company 94
Corporation of Barbados Adventurors
    58
Cortés, Hernán 24–5
cotton 49, 51, 87; from Brazil 89; from
    India 84; from North America
    136–7, 151; from Spanish America
    90
cotton gin 136
Courland 55
Courteen, William 49
credit: in African slavery 100; in
    Atlantic slave trade 33, 51, 64, 84;
    and triangular trade 92
creole language 71
creolization 70
Crimean Tartars 170
Crittenden, John 153
Cromwell, Oliver 4
Crusades 12, 23

Cuba ix; British capture of 86, 131; convict labor in 4; end of slavery in 151, 156; indentured labor in 163; slavery in 28, 70, 134–5, 138–9, 147; Spanish conquest of 25
cultural practices 3, 70
cultural relativism 111
Curaçao 53; British capture of 86; creole culture in 71; as slave entrepôt 45, 48, 68, 80; slave revolt in 128–9

Dahomey 55, 63, 97–100
Danish West Indies 81, 86, 120, 132, 150
Danton, Georges 127
Darfur 9, 14–5, 144, 182
Dartmouth 92
Davis, Garrett 128
Day, Thomas 111
de la Cosa, Juan 39
de la Founte, William 37
death rates: in American colonies 56–7; in Atlantic slave trade 33, 66–8, 97
deaths in passage 2
debt: repayment of 60, 64–5; slaves as collateral for 52
debt bondage 3, 180
Delaware 151, 153
Delos 9–10
Demerara-Essequibo 48, 132, 145–6
Democratic Party 152, 154
demographic self-sufficiency 105–6, 145
Denmark 150, 181; abolition of slavery in 120; and slavery 80–1; in West Africa 53, 55, 63
Dent, William 125
dependency 4, 33
deportation 4, 121–2
Dessalines, Jean Jacques 128
Dias, Bartolomeu 36
Diderot, Denis 113
Dolben Act 123–4
domestic servitude x, 9, 32, 37, 59
Dominica 48, 86, 105, 128
Dominican Republic 24, 127
Douglas, Stephen 152
Drake, Francis 38
Dred Scott decision 151–2
Du Bois, William 177
Duncan, Edward 141
Dunmore, Earl of 112
Dutch Reformed Church 126

Dutch West Africa Company 53
Dutch West India Company 64, 96
dyewoods 65, 93

Eanes, Gil 31
East Africa 2; Portugal in 62, 99; slave trade in 15, 41, 82, 166–7
East Indies 22, 82, 92, 143
East Indies Squadron 167
Eastern Europe 15–7, 34
economic opportunity 4, 8, 41
Ecuador 90
Egypt: expansionism in north-east Africa 144; Ottoman conquest of 14, 22; slavery in 6, 9–10
Eisenhower, Dwight 177
Ejiofor, Chiwetel xi
Elizabeth I of England 37
Elliot, Charles 137
Elliott, Gilbert 142
Elmina 30–1, 36, 54, 89
Emancipation Act 1833 147–8
Emancipation Proclamation 154, 177
encomienda 26, 46
Encyclopédie 113
England: Anglo-Saxon 11; Atlantic colonies of 49; legality of slavery in 13, 121; pioneers of slavery 37–8; in West Africa 53–4; *see also* Britain
English Civil War 53
English slave trade 56–60; *see also* Britain, and the slave trade
English West Indies *see* British West Indies
Enlightenment 76–7, 110–11, 113
entrepôts 41, 48, 68, 80, 85, 89
equality, legal 17, 154, 164
Equatorial Guinea 144
Eritrea 9–10, 166
escaped slaves *see* fugitive slaves
Essequibo 48, 132
Ethiopia: in 16th century 14, 34; Portuguese in 39
Euro-African society 65
Europe: Christian 12, 20n21, 24; medieval 3, 11
European Union 183
Eusébio de Queiroz law 139
Evangelical Christianity 79, 113, 123, 130, 133
Exeter 92

Exmouth, Lord 140
Eyre, Edward 165

Factory Act 1833 148–9
families of slaves 59–60
Farragut, David 143
female slaves 59, 65, 105
Fernando Poo 36, 87
fertility of slaves 105
Fielding, Henry 7
filibusters 137–8
finance, international 44–5, 87–8, 92, 134
Financial Revolution 80
firearms 15; in Africa 35, 62, 98–9, 135
Florida: French in 38; slave flight to
　101, 103, 146; Spanish in 25, 79
forced labor x; in the Gulf States 184; in
　Madagascar 16; from Native
　Americans 37; serfdom as 11; and
　slavery 3, 41; in World War II 170–1
forced marriage x, 3
Foreign Slave Trade Act 1806 131
former slaves viii–ix, 69–71, 127, 148,
　162–3; states for 168
Formidable 140
Fort James 54, 98
Fort Jesus 22, 62
Fort Louis 98, 132
Fort Mose 101–2
Fort Sumter 153
Fort Wagner 163
Founding Fathers 164, 179–80
Fox, Charles James 131
Fox, Henry 7
Fox, William 124
France 7; abolition of slavery in 76,
　126–9, 150–1; African empire of 98,
　144; Atlantic empire of 48–9; in
　Madagascar 16; and the slave trade
　56, 79–82, 133, 138, 150; Wars of
　Religion in 38; wars with Britain
　85–6, 102, 129, 140
Francis, Pope x
Franco-Dutch war 48, 52, 56
free Africans 70, 168
free blacks: in Brazil 155; colonies for
　141; in French Guiana 90; in
　Jamaica 105; in Spanish Florida
　101–2; in Texas 137
free labor 142, 147, 149, 155–6, 165
Free Soil Party 152

free trade 56, 80, 89, 147, 149, 164
freed slaves 70, 75, 141, 168; diaspora
　of 112
Freeman, William 46
French Guiana 48, 90
French Guinea Company 79
French Revolution 76, 85–6, 110, 112,
　119, 125–7, 129
French West Indies 82–3, 85–7, 103;
　economic value of 94, 127;
　indentured labor in 163
Frostrup, Mariella 180
fugitive slaves 21, 52; advertising about
　113; in American South 152; in
　Brazil 156; organized communities
　of 102–3; recapture and deportation
　of 121–2; and Seminole Wars 146

Gabon 144, 168
galley slaves 6, 78
Gambia 85, 98, 174, 182
Gambia River 31, 34, 53, 100
Gangmasters Licensing Authority ix–x
Gastaldi, Giacomo 39
gender 36, 65, 68, 106, 177, 180, 186
geopolitics 2, 22, 31, 54
George III of Great Britain 7, 86,
　111–12, 122, 124, 131, 179
Georgia: in American War of
　Independence 86; rice cultivation in
　70, 91; slave imports 112; slavery in
　106, 110; slaves fleeing 103; Spanish
　attack on 102
Germany 62, 167, 170–1
Ghana 53–4, 63, 177
ghettos 171
Gibbon, Edward 122
Gillray, James 124
ginger 51
Glasgow 92, 94, 113
Glehue 63, 99
globalization ix, 17
Glorious Revolution 56, 80
gold: from Brazil 88, 134; and
　Portuguese in Africa 31–2
Gold Coast 31; cooperation with local
　rulers 64; Denmark and 81; Dutch in
　80; England in 53–4; impact of slave
　trade on 60–1; Portugal in 19;
　warfare in 63
Golden Horde 23

Golden Law 156
gold-slave economy 32, 88
Gordon, George 165
Gorée 49, 79, 82, 86, 98
Granada 15, 18
Grau, Justin 176
Great Exhibition in London 1851 149
Great Reform Act 1832 147–8
Greece, ancient 9
Grenada 48–9, 86, 105, 128
Grenville, Lord 131–2
Grey, Earl 147
Guadeloupe 22, 48, 53; British capture
  of 86, 131–2; cacao in 46; restoration
  of slavery in 127; slave uprising in
  150; slaves sent to 82
Guanches 24, 30
Guatemala 25
Guianas 48, 145
guilt, hereditary 186
Guinea: Portugal in 19; slaving
  networks in 47
Guinea Company 53
gulags *see* labor camps
gum arabic 101
gun-slave cycle 60
Gustavus III of Sweden 81
Guyana 106, 132, 146, 163–4

Hackwood, William 123
Haiti 24, 48
Haitian Revolution 127–8
Hamburg 83, 85
Hausa states 99
Hawkesbury, Lord 124
Hawkins, John 37, 174
Heming, Sally 180
Henry VII of England 37
Hispaniola 24, 30, 32, 37, 53
HMS *Buzzard* 140
HMS *London* 167
HMS *Menai* 141
HMS *Monkey* 141
HMS *Pickle* 141
HMS *Sampson* 142
Hogarth, William 111
Hollandia 80–1
Holmes, Robert 53
Holocaust xi, 101, 170, 187
Holt, John 13
homeless 183

Honduras 25–7
Horn of Africa 14, 34, 144
House of Commons 124, 146
House of Lords 124, 131
household service *see* domestic
  servitude
huckstering 70
Huggins, William 141
human rights 56, 126

Ibo 106
Iceland 11, 34
identity politics 185–6
Idris Aloma 14
indentured labor: in British Empire
  163–4, 167; in the Gulf States 184;
  and slavery ix, 4, 41, 50–2, 107; and
  sugar 46; and tobacco 58
India: British Empire in 122; early
  Western bases in 22, 62–3; French in
  82; indentured labor from 163–4;
  maritime pathway to 36; Portuguese
  90; and sugar 84
Indian Ocean: British trade in 93;
  French power in 82; Portuguese in
  36, 62; slave trade across 13, 15–6,
  22, 65, 100
indigo 49, 51, 59, 83, 90
Industrial Revolution 60, 113
industrialization 113, 151, 156
infectious disease 24, 67–8
inheritance of slaves 59
intermarriage 71, 168
International Association Against
  Slavery 3
International Convention with the
  Object of Securing the Abolition of
  Slavery and the Slave Trade 2
International Slavery Museum,
  Liverpool 174
interracial sex, consensual 59, 106
Ireland 4, 11, 130
Isabella of Castille 26
Islam: conflict with Portugal and Spain
  18; and slavery 5, 12, 15, 66
Islamic world: mapping in 39; slavery in
  6, 8, 12, 15, 41, 166
Ivory Coast 82, 143–4

Jackson, Andrew 133
Jackson, Jesse Jr. 178

Jacobites 6–7
Jamaica ix; cacao in 46; economic
    pattern of 51; Morant Bay uprising
    in 165; naval yards at 91; and
    Panama Canal 163; runaway slaves
    in 52, 102–3; settlement of 50; as
    slave entrepôt 68; slave life in 105–6;
    slave rebellions in 69, 102, 146–7;
    women and children as slaves in 15
Jamaica Slave Act 52, 59, 69
James I of England and VI of Scotland
    45, 53
James II of England 53, 56
Jamestown 57
janissaries 5
Japanese imperialism 170, 176, 185
Java 17
Jefferson, Thomas 110, 128, 133,
    179–80
Jerba 14, 18
Jesuits 89, 107
Jesup, Thomas 146
Jewish Naturalization Act 126
Jews: in Exodus narrative 7; and
    hereditary guilt 186; in Nazi
    Germany xi, 101, 170–1
jihad 14, 66, 99
Johnson, Andrew 164
Julia de Roubigné (Mackenzie) 78
July Monarchy 150

kafala system 184
Kano 62, 166, 183
Kansas-Nebraska Act 152
Kendal 94
Kentucky 128, 151, 153–4
Kilwa 100, 166
King, Martin Luther Jr 174, 177
King's Negroes 91
Kongo 36, 97
Ku Klux Klan 164

La Rochelle 94
labor camps 4, 170–1
labor control 11–2, 16–8, 162–3,
    167, 187
labor flows x–xi, 41, 163, 171, 182–3
labor regimes 46, 51
Lacerda e Almeida, Francisco 100
Lagos 142
Laird, Macgregor 142

Lancaster 92–4
Latin America 1, 24–5; abolition of
    slavery in 138, 151; demand for
    slaves in 79; intermarriage in 71;
    labor flows from 183
latitude, measuring 39
Le Cat, Claude Nicolas 108
Le Rodeur 119
Leclerc, Charles 128
Leo Africanus 40
Liberia 66, 128, 168
liberty, and slavery 6–7
Libya 182
Lincoln (film) 178
Lincoln, Abraham 151–4, 178, 180
Lisbon 32, 36, 38
literature: black people in 109; and
    slavery 76–8, 123
Liverpool 44, 90–2, 109, 136, 174
Livingstone, David 167
Locke, John 59
London: anti-slavery demonstrations in
    175–6; and international finance 80,
    88; as slave port 90, 92
Louis XVIII of France 133
Louisiana 48; French loss of 150; slave
    rebellion in 146; slaves sent to 82
L'Ouverture, Toussaint 127–8
Lower California 137
Luanda 30, 47, 87, 89
Lunda Empire 97, 99
Luso-African families 89
Lyons, Lord 152

Maccarthy, Charles 144
Macedon 9–10
MacHardy, J.B.B. 141
Mackau laws 150
Mackenzie, Henry 78
McQueen, Steve xi
Madagascar 15–6, 63, 82, 126
Madeira 23, 31–2
Madison, James 180
Mahi 98
mahogany 84, 93
malaria 57, 142
Malcolm X 177
Malcontents 110
male labor 15, 65
male slaves 36, 59
Mali 18, 31, 33, 60, 66, 98, 182, 186

Malinke 106
malnutrition 67, 142
Malpighi, Marcello 69
mamelucos 89
Mamluks 6, 21–2
Manchester 136
Mansa Munsa 31
Mansfield, Lord 121
mapping 28–9, 31, 38–40, 130
Maroons 21, 102–3, 128
Marquês de Pombal 88
martial law 102, 165
Martinique 22, 48, 52–3; British capture of 86, 131–2; cacao in 46; restoration of slavery in 127; slave uprisings in 150; slaves sent to 82
Mary I of England 37
Maryland 49, 57, 102, 137, 151, 153
Massachusetts 112
Mauritania 169, 182
Mauritius 15, 22, 82, 134, 145
May, Theresa x
Mayan civilization 25–6
measles 30, 89
memorialization 175
memory, public 174
Mercator projection 29, 40
Mercier, Louis Sébastien 75–6
mestizos 30
Methodists 120–1, 123
Mexico 10, 35; abolition of slavery in 138; caste system in 26; colonial demographics of 30; Spanish conquest of 24–5
Mfecane 186
Midas 141
Middle Passage 1, 67–8
military history 9, 181
Miller, Joseph 67, 189
Milward, Anthony William 140
Minas Gerais 87–9
mining 9
Minnesota 151
Mississippi 49, 178
Missouri Compromise 137, 151, 153
Mombasa 22, 62, 99–100
Mongol Empire 16, 23
monogenesis 110, 167
Montserrat 46, 49, 86, 102
Moors: as slaves 23, 27, 41; warfare with Christians 12, 18, 24

moral corruption 59–60
Morkunas, Audrias x
Morocco: and Portugal 27, 37; slave soldiers in 6; and slave trade 13–4, 33–4
mosquitoes 57
movement, compulsory 3–4
Mozambique 61–3, 90–1, 100, 126, 134
Mughal Empire 13, 16, 22
Muhammad Yusuf, Sultan 62
mulattos 71, 90, 105, 129
muskets 62–3, 98
mutilation 66, 150, 180

Napoleon Bonaparte 127
Napoleonic Wars 112, 121, 130–2
Nasr al-Din 66
Nation of Islam 177
National Association for the Advancement of Colored People 177
Native Americans 41n4; and African slaves 28–30, 33; contemporary conditions of 180; in *encomienda* system 46; forced labor of 37, 139; in French Guiana 90; slave raiding by 25–6; as slaves 58, 87, 89, 109; Spanish allies 101–2
naval power 49, 56, 91, 120, 140, 183
naval yards 91
navigational expertise 39
Nazi regime 170
Negroes, as legal category 13, 51–2
Netherlands: abolition of slave trade 133; and Atlantic slave trade 37–8; and Brazil 47–8; historical memory of slavery 181; and the slave trade 80; in Taiwan 22; war of independence 38; in West Africa 53–4
Nevis 49, 86, 105
New England 84
New York, abolition of slavery in 112
Newfoundland 17, 44, 94
Nicaragua 26–7, 137–8
Nichols, Vincent x
Niger 66
Niger Delta 60
Niger River 40, 70, 97, 142–4
Niger Valley 97
Nigeria 62, 66, 99, 166, 169, 186
Nixon, Richard 177
Nonconformists 148

North Africa 10; Portugal and Spain in 23, 27; slave trade in 14–5, 18
North America 2, 27; British/English 49, 57–9, 90–2, 105–6; slaves born in 102; *see also* American South
North Atlantic 46–7, 89, 95, 139
North Carolina 84
North Korea xi, 6, 170–1, 183–4, 186
Nova Scotia 112

Obama, Barack 79, 178
Ohio River country 85
Oman 41, 62, 99, 166–7
Opubu the Great 135
Orders-in-Council 131–3
Oregon 151
Organization Todt 171
origin myths 8
the Other, and slavery 8, 12–3
Ottoman Empire 5–6; coffee in 83; expansion of 21–3
Owojaiye, Matthew 169
Oyo 63, 99

palm oil 135, 168, 183
Palmerston Act 138
pamphlet literature 123
Panama 27, 163
pardos 30
Park, Mungo 100
Paulus, Lucius Aemilius 9
pawning 65, 168
Pennant, Thomas 94
Pennsylvania 104, 120, 152
people smuggling 4
Peregrine Pickle (Smollett) 77–8
Peru 25–7, 29, 35, 90, 138
Philip II of Spain 37–8
Philip V of Spain 79
pioneers 91
Piper, N. A. 184
piracy 10, 141–2
Pitt, William the Younger 122–5, 131
plantation economies 15; in Africa and Arabia 165–7; around Indian Ocean 22; in Atlantic world 23, 59, 64; development of 83; on Java 17; in North America 86, 91; in Portuguese possessions 32; and slave trade 55; in West Indies 49–52, 129, 145, 164–5; *see also* sugar plantations

plantation goods 55, 82–4, 89, 92
plantations, historical 180
Plessy v. Ferguson 164
Plymouth 92
Poland 170–1
political arithmetic 80
polygamy 65
polygenism 108, 167
Poole 92
Poor Law Amendment Act 1834 148–9
Portsmouth 92
Portugal: and African slaves 18–9, 31–2; Atlantic conquests of 23, 25; and British abolitionism 132–3; deporting convicts 4; economic relationship with Brazil 88; end of slave trade 138; ending native slavery 26; expansion into Africa 13–4, 31, 34–5, 61–2; slave economy of 87; in South Atlantic 47–8; union with Spain 37–8; *see also* Brazil
Portuguese Guinea 87
Postlethwayt, Malachy 121
Principe 87
prisoners of war, enslavement of 8–9, 16
prostitution, forced ix–x, 3, 176, 183–4
protectionism 93, 121, 129, 164, 183
Protestantism 17, 119–20, 122, 131, 150
providentialism 130
public space 80, 84
Puerto Rico 30, 136, 143
purchasing power 64

Qatar 184
Quakers 104, 119–20, 122

racial classification 40, 69, 108, 110, 138
racial hierarchy 6, 69, 145
racial identity 59
racial mixing 71, 138
racial typecasting 27–8, 40–1
racism 1; in Brazil 155; in British West Indies 146, 165; and slavery 8–11, 17, 41, 51, 57–8, 69, 107–11, 124
Radishchev, Alexander 3
raiding 169; in Africa 14, 18, 33–4, 166; in the Americas 25–6, 30; in Brazil 89; British action against 141; by Golden Horde 23
rape 6, 66, 106, 180
rationalism 146

rationality 40
Raynal, Abbé 113
Recife 47
Reconquista 18, 25
Reconstruction Acts 164
Red Sea 10, 13, 15–6, 143, 166
reducciones 89
reparations for slavery viii, xi, 185–7
repartimiento 26
Republican Party 152–4, 164, 178
Réunion 15, 82, 134
Revolt of the Tailors 129
Rhode Island 94
rice cultivation 46, 59, 70, 105
rigging 39
Rio Branco law 155
Rio de Janeiro 38, 89, 156
rip-off work 183
Robertson, William 111
Robespierre, Maximilien 126–7
Roderick Random (Smollett) 77–8
Roman world 5, 9–10, 77; and British
    Empire 122
Romero, George 185
Rousseau, Jean-Jacques 111
Rowson, Susanna 136
Royal African Company 35, 56, 63–4;
    Colston and 175; deaths of slaves
    traded by 67; end of 90, 95–6; and
    South Sea Company 80
RSPCA 130
rum 93–4, 105, 126
runaway slaves *see* fugitive slaves
Russia: in Central Asia 169; serfdom in
    3, 17, 163; southern 10, 23

Sahara, slave trade across 10, 13–6, 33–4
sahel 14, 33, 64, 97, 140, 166
Saint-Domingue 53; slave population of
    82; slave rebellion in 85, 102, 113,
    127–8; sugar from 83; Vodou in 103;
    see also Haiti
sale of slaves 25, 31, 33, 64–6
Salé Rovers 34
San Basilio de Palenque 21
São Paulo 30, 87, 89, 136, 155
São Tomé 32, 47, 87
Saudi Arabia 6, 169, 184
Schoekher, Victor 150
Scientific Revolution 21, 29, 69, 76
Scipio Africanus 9

Scotland 11, 45, 94, 113; *see also* Britain
Scottish Guinea Company 53
Sebastian I of Portugal 14, 37
segregation 164, 177
self-respect 104
Seminole Wars 146
Senegal 79, 86, 98, 182
Senegal River 31, 34, 66, 82, 98, 132
Senegal River Valley 98
Senegambia 18, 31, 47, 60, 86, 106
serfdom 3–4, 11–2; second 16–7, 34
Seven Years' War 85–6
sexual exploitation of slaves x, 6, 15, 59,
    65, 105–6, 169, 183
Shaftesbury, Earl of 58–9
Shakespeare, William 12, 49, 176, 187
Shama 30, 36, 80
Shang dynasty 4
Sharp, Granville 104
Shaw, Thomas 140
Sherer, Joseph 141
Sherman, William T. 163
Shettima, Kashim 169
shipbuilding 38–40, 106
Siberia 17
Sicily 13, 18, 182
Sierra Leone 36, 54, 66, 112, 141, 168
silver 29–30
Simon, Herbert 184
1619 Project 179
skills of slaves 70
skin color 12, 40, 108
Slave Coast 63, 86, 97
slave economy: British 91, 102, 129;
    coercive nature of 104; dependency
    on Africans 70; and disease 57; in
    North America 59; and plantation
    economy 52; Portuguese-West
    African 32
slave flows 65, 79, 106
slave labor 22–3; in Brazil 32, 47;
    demand for 64–5, 134; in North
    America 56–8; in South Africa 63; in
    West Indies 51
slave life, in Americas 6, 69–71, 104–7
slave markets 14, 18, 32, 144; in West
    Indies 51
slave ownership: compensation for 148;
    sequential 67; UCL database of 95
slave rebellions 85–6, 96, 101–3, 127–9,
    139, 146, 151

slave ships 34, 66; naval action against 138–43; regulating conditions on 123–4; risings on 101; size of 95; source of 92

slave societies 4–5; mature 145

slave soldiers 5–6, 166; *see also* black soldiers

slave trade: in antiquity 9–10; and the Enlightenment 76; and geopolitics 21–2; and illegal activity 50; major centers of 41; modern ix–xi, 182–5; multiple 2; non-Western 13; in Ottoman Empire 23; significance of viii–ix; *see also* African slave trades; Atlantic slave trade

Slave Trade Act 1845 139

slavery: arguments in favor of 108–9; definitions of x, 2–8, 41; further reading on 189–90; legacies of 174–82; and the Other 12–3; and racism 8–11; and serfdom 11–2; in twentieth century 169–72

smallpox 24, 30, 89

Smith, Adam 4, 95

Smollett, Tobias 77–8

smuggling 48, 50; of slaves 134

Snelgrave, William 109

Social Darwinism 145

social protests 103

social stratification 33, 51

Société des Amis des Noirs 126

Society for Effecting the Abolition of the Slave Trade 122–3

Society for the Extinction of the Slave Trade and the Civilization of Africa 142

Society of West India Merchants and Planters 124

Sokoto Caliphate 99, 166, 169

Somerset, James 121

Songhai Empire 13, 33–4

Sonthonax, Léger 127

Souk el Berka 140

South Africa 63

South Atlantic 36, 46–8, 87, 89, 139

South Carolina 49, 69; and Confederate flag 179; rice cultivation in 70, 91; secession of 153; slave imports 112; slave population of 105–6; slave rebellions in 101, 110, 146; slaves fleeing 103

South Sea Bubble 80

South Sudan 9, 15, 186

South-East Africa 61, 90, 93, 134

Southern identity 2, 151

Soviet Union 4, 170

Spain: in Africa 36, 144; Atlantic conquests of 23–5; conflict with England 37; deporting convicts 4; Moorish 15; restricting trade with colonies 93; *see also* Aragon; Castile

Spanish America: African slavery in 27–30, 33, 37, 89–90; British attacks on 84–5; independence of 138; sale of slaves to 79–80; tied labor in 26, 46; urban slaves in 71; *see also* Latin America

Spanish West Indies 89

SPG (Society for the Propagation of the Gospel in Foreign Parts) 107–8

St. Augustine, Florida 25, 101, 103

St. Barthélemy 81, 151

St. Christopher *see* St. Kitts

St. Eustatius 48, 80, 132

St. Kitts 48, 86

St. Louis, Senegal 49, 82, 86

St. Lucia 49, 86, 132, 150

St. Martin 132

St. Mary Redcliff Church 176

St. Vincent 86, 105, 128

stadial theory of change 75

Stalin, Joseph 4

Stanley, Henry 144

state slavery 5–6, 91, 170–1, 185–6

steam power 141–2, 144, 165, 183

Stephens, James 133

Stevenson, Charles 143

Stono Rebellion 101–2, 110

sub-Saharan Africa 6; European expansionism in 63; human trafficking from 169; Moroccan expansionism in 13–4; Portuguese and Spaniards in 18; Western slave trade with 10, 33–4, 40–1

Sudan x, 186; and ancient slavery 9–10; modern warfare in 66, 169; raiding in 144, 166; warfare in 66

sugar 44–6; boycott of 122, 124; British sources of 164; Danish interest in 81; direct export to Europe 96; European consumption of 83–4; French sources of 82

sugar cane 13, 30, 46
Sugar Duties Act 1846 164
sugar plantations: African slaves on 32;
  Guanches on 30; Irish prisoners on 4
sugar production 46; after slavery
  162–3; in Brazil 30, 32, 47–8, 87,
  165; in British/English West Indies
  49–51, 105; in Cuba 147; and disease
  57; in French West Indies 83; in
  Mauritius 22; in Spanish America 90
suicides 102, 105, 147
Sulivan, George 167
Surinam 46, 48; British capture of 86; in
  Candide 76, 78, 104; end of slavery
  in 151; fugitive slaves in 103
Swahili Coast 62
Sweden 53, 55, 81, 150
Swedish African Company 64
Swedish West Indies 86
Sydney, Viscount 124
syncretism 36, 71, 177

Taînos 30
Taiwan 22
Tallmadge, James 137
tea 22, 44–5, 88
The Tempest 12
Tenison, Thomas 107
terms of trade 98
Texas 137
textiles 93–4, 134–5, 166, 182–3
Thailand 184
Thistlewood, Arthur 7
Thistlewood, Thomas 105
Thurlow, Lord 124
Timbuktu 14, 97, 143
tobacco 44, 46; African demand for 55;
  in Brazil 89, 99; European
  consumption of 82–3; in North
  America 57–8, 106; in Spanish
  America 90; in West Indies 49
Tobago 85–6, 102, 105, 128, 132, 150
Topsham 92
trafficking x, 4, 183–4
transnationalism ix
Treaty of Alcáçovas 19, 36
Treaty of Saragosa 36
Treaty of Tordesillas 36
triangular trade 92–4
tribal identity 34
Trinidad 49; British capture of 86, 145

Tripoli 14, 32, 140
tropical disease 57, 97
tropical goods 84, 130
tropical lands, Western typecasting
  of 28
Trustees 110
Tuckey, James 144
Tunis 14, 32, 140, 144
Turnbull, David 139
Turner, Charles 144
Turner, J.M.W. 75
Turner, Nat 137
Tuscany 55
Twelve Years a Slave xi

Uganda 66
United States Constitution 112;
  Thirteenth Amendment 154,
  164, 178
United States of America: abolition of
  slave trade 133; after slavery 163–4;
  end of slavery in 112, 151–4; legacy
  of slavery in 177–80; naval attacks
  on slave trade 142–3; and reparations
  186; slavery in 2, 135–8; *see also*
  American Civil War; American
  South; American War of
  Independence
universalism, secular 113

Venezuela 48, 89–90; abolition of
  slavery in 138; cacao from 45, 83; slave
  risings in 103, 129
Veronese, Paolo 32
Vespucci, Juan 39
Villa, Monique viii, x
Virginia 49; in American War of
  Independence 86; black governor of
  178; secession of 153; slave rebellions
  in 137, 146; tobacco in 57–8
Vodou 103
Volodora 141
Voltaire, François Marie Arouet de 76,
  79, 104

Wadström, Carl Bernhard 81
wage slavery 3
Walker, Adam 113
Walker, William 137–8
Walpole, Robert 107
War of 1812 133

War of American Independence *see* American Revolution
War of Jenkins' Ear 84, 110
War of the Austrian Succession 7, 85
War of the Spanish Succession 80
Wedgwood, Josiah 123
Wesley, Charles 104
Wesley, John 104, 120–1
West Africa x, 10; and Atlantic slave trade 30, 47, 53–5, 61, 95, 182; conflicts of Western powers in 55, 63; cooperation with local rulers 64–5; end of slave trade in 166; English and British in 37–8, 85, 98, 140–2; European typecasting of 28; female slaves in 65; military developments in 62–3, 98–9; non-Western slave trades in 13–4; patterns of trade in 101; Portuguese in 31–3, 35–6; slave ports of 67; Spanish absence from 36–7; warfare in 146
West Indies 1, 15, 21; and Atlantic slave trade 48–54, 81; coffee from 22, 84; indentured labor in 4; native population of 27; sugar production in 47; *see also* British West Indies; Danish West Indies; French West Indies; Spanish West Indies; Swedish West Indies
West Virginia 153
Western, use of term 20n21
Western Design, English 53

Western expansion 17–8, 22–3, 38
Western slavery 8, 16–7, 22
Whigs (Britain) 7, 13, 59, 147
Whigs (USA) 152
whipping 52, 78, 108, 180
white slaves 12
Whitehaven 92, 96
whiteness 41
Whitney, Eli 136
Whydah 54, 85, 99
Wilberforce, William 121, 124–5, 131–3, 145, 180
Wilder, Douglas 178
William III of England 56
William IV of the UK 124
women: consequences of slavery for ix; white 106; *see also* female slaves; gender
workhouses 149, 162
Wright, Joseph 109

Xhosa 99

yellow fever 56–7, 128, 141
Yemoja 71
Yoruba religion 71
Yucatán 25–6, 143

Zambezi River 61, 100
Zanzibar 100, 166–7
Zollverein 121
Zong massacre 75, 97